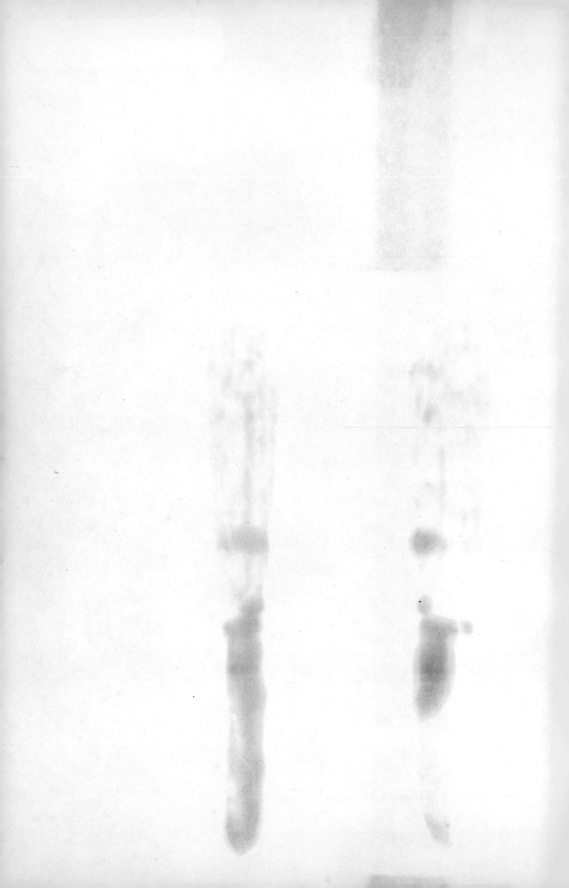

Nim

VIII

Nim

Herbert S. Terrace

Alfred A. Knopf   New York   1979

THIS IS A BORZOI BOOK
PUBLISHED BY ALFRED A. KNOPF, INC.

Grateful acknowledgment is made to H. W. Hoemann and the National Association of the Deaf for permission to reprint excerpts from *The American Sign Language* (1976). Reproduced with the permission of the author, H. W. Hoemann, and the publisher, the National Association of the Deaf, 814 Thayer Avenue, Silver Spring, Md.

Acknowledgment is also made to the following individuals for permission to reproduce the photographs on the pages specified: Robert Bazell, 230; Harry Benson, 54; Connie Garlock, 36; Susan Kuklin, 78, 91, 105, 164 (top right, bottom left and right), 195, 200; Penelope Lockridge, 51, 146; Laura Petitto, 177; and Gerald A. Tate, 77, 148 (top left, center left and right, bottom right), 151, 160, 199, 201, 205, 206, 243 (bottom right).

Library of Congress Cataloging in Publication Data

Terrace, Herbert S   [date]
    Nim.

Includes index.
1. Chimpanzees—Psychology. 2. Human-animal communication. 3. Sign language. 4. Mammals—Psychology. I. Title.
QL737.P96T47 1979   156'.3   79-2157
ISBN 0-394-40250-2

Manufactured in the United States of America
FIRST EDITION

To Laura Ann Petitto,
an inspiration to Nim and to me

Contents

Acknowledgments

This book is an informal account of an effort to communicate with a member of another species via a human language. The goal was to teach American Sign Language to an infant chimpanzee, Nim Chimpsky, and to document the conditions under which Nim communicated in this medium. Unlike most scientific projects that can be conducted by one person (so long as that person is patient enough), this project could only be conducted as a group effort. The most energetic and organized person imaginable would have neither the time nor the stamina to look after a growing chimpanzee, to teach it to sign, and to record and analyze the volumes of data needed to demonstrate just what Nim learned about sign language.

During the four years of the project, Nim was looked after and taught by sixty teachers. Most of these teachers were volunteers. Another group of approximately forty volunteers tabulated and analyzed data from daily records and videotapes. Each of the more than one hundred project members made valuable contributions to the project, often at considerable personal sacrifice. The special efforts of three individuals significantly advanced both the growth of the project and Nim's social and intellectual development. I thank Stephanie LaFarge, Laura Petitto, and Bill Tynan for their many unique contributions. Other project members deserve special mention for the quality of their work and the conscientiousness they showed in spending extra hours either teaching Nim or analyzing data: Walter Benesch, Ilene Brody, Joyce Butler, Bob Johnson, Andrea Liebert, Susan Quinby, Dick Sanders, and Amy Schachter. I am also grateful to Dr. Buddy Stark, Steve Lerman, and Tom Blumenfeld for the splendid, and often free, medical care they gave to Nim.

Nim's existence as a special student receiving intensive instruction in sign language also required material support. That was provided by Columbia University and various foundations. Professor Julian Hochberg, who was chairman of Columbia's Department of Psychology when the project began, provided a much needed loan of video equipment and funds for the minimal salary of Nim's one full-time teacher. Dr. William J. McGill, and Mr. William Bloor, Columbia University's president and treasurer respectively, kindly arranged for the use of the Delafield estate, Nim's home for the last two years of the project. Grants from the W. T. Grant Foundation, the Harry Frank Guggenheim Foundation, and the National Institute of Mental Health made it possible for the project to obtain valuable data on Nim's use of sign language and to function as long as it did. W. E. R. LaFarge generously made available his house as a center for Nim's upbringing and instruction during the first twenty months of Project Nim. Professor Tom Bever, a colleague in Columbia's Department of Psychology, argued persuasively in a joint effort to obtain funding from these foundations. The Children's Television Workshop, through the efforts of Edith Zarnow, provided crucial financial assistance in return for which various aspects of Nim's social development were filmed for use on the Sesame Street program.

Professor Lois Bloom of Teachers College furnished much helpful information about language learning by children. A wise and understanding editor, Charles Elliott, gave me valuable advice and encouragement in writing this book. I am also grateful for the comments of friends who read a number of portions of the book in various stages of its evolution: Julie Baumgold, Susan Braudy, Marnie Long, Laura Petitto, and Jeffrey Steingarten. Susan Kuklin, a friend and professional photographer who contributed some of the photographs that appear in this book, proved her friendship in nonphotographic roles as well: her mature presence often ameliorated personal tensions between project members that I did not have time to deal with myself. Special thanks are extended to Gerald A. Tate for his invaluable assistance in processing and printing many of the photographs. Last, but hardly least, I would like to thank Dr. W. Lemmon, director of the Institute of Primate Studies, for providing me, at no cost, with a healthy newborn male chimpanzee.

During the course of this project I learned a great deal about how a chimpanzee can communicate through the medium of a human language. I also learned a great deal about how some wonderful people found a way to work together to create the special environment needed to foster and document communication with a member of another species. To each of them I extend my warm and heartfelt thanks.

Nim

1

Introduction

Consider the following conversation.

Teacher:	*What want you?*
Student:	*Eat more apple.*
Teacher:	*Who want eat more apple?*
Student:	*Me Nim eat more apple.*
Teacher:	*What color apple?*
Student:	*Apple red.*
Teacher:	*Want you more eat?*
Student:	*Banana, raisin.*

A short time later, the teacher and the student are taking a walk. This time, the student starts the conversation.

Student:	*Bird there.*
Teacher:	*Who there?*
Student:	*Bird.* (pause; looks in other direction) *Bug, flower there.*
Teacher:	*Yes, many things see.*
Student:	(rolls over on ground) *You tickle me.*
Teacher:	*Where?*
Student:	*Here* (pointing to leg).
Teacher:	(after tickling) *Now you tickle me.*
Student:	(tickles teacher) *Me tickle Laura.*

If the teacher and student had been speaking aloud, this is how their conversations might have sounded, like the "baby talk" that often occurs between a young child and an adult. But the "student" was a two-and-a-half-year-old chimpanzee named Nim, and the conversations were conducted in utter silence. All communication was based on American Sign Language, a complex natural language of hand gestures and facial expressions used by thousands of deaf people in the United States.*

---

* See Appendix A for information about the differences between sign and spoken languages.

This book is a personal account of a scientific project whose main goal was to teach an infant chimpanzee to use language. There were many reasons for starting the project, but the most compelling for me was that it might define more clearly what it means to be human. Language and the culture to which it is essential have long been considered the exclusive property of humans. If indeed a chimpanzee could learn to use language in a humanlike manner, the distinction between humans and at least one nonhuman animal would have to be reexamined.

I also wanted to learn, through direct conversation, how another species perceives the world. Previously, this question had been considered only in the realm of science fiction. Nevertheless, we approach the twenty-first century with the serious expectation of discovering life elsewhere in our universe. If extraterrestrial life is discovered, one of the first things we will ask is whether we can communicate with such beings. Yet ironically, at the dawn of the age of space, our first opportunity to engage in intelligent communication with a nonhuman species is available right here on earth. A chimpanzee may not be as exotic as a creature from outer space, but communicating with any nonhuman presents the same romantic prospect, at once fascinating and confounding, of attempting to understand a point of view wholly outside our experience. It also poses a now-controversial question that linguists, psychologists, psycholinguists, philosophers, and others have yet to resolve: What is distinctive about human language?

Until recently, humans could take comfort in the assurance that our language made us unique. This seemed clear no matter how one defined the distinctive features of human languages. Now, the unexpected success of several attempts to teach elements of language to chimpanzees has created new doubts about our distinctiveness and given rise to pressure to define human language more clearly. How else can we assess the magnitude of an ape's intrusion into what has previously been regarded as an exclusively human preserve?

One reaction has been to define human language in a way that would automatically exclude the achievements of the two dozen or so chimpanzees that appear to have mastered some of its features. In other words, if a chimpanzee can learn it, it isn't language. The languages taught to the chimpanzees were in the form of gestures or artificial plastic symbols. By insisting that language be spoken, it is easy to deny that chimpanzees can use language. But this kind of arbitrariness runs into one major difficulty: American Sign Language, whose signs have been learned by at least a dozen chimpanzees, is but one of a family of nonvocal languages used by hundreds of thousands of deaf people around the world. Even if none of the languages used by chimpanzees had a direct human analog, it might still be possible to teach a chimpanzee the essential features of human language, whatever those features might be. Of course, one could keep refining what a chimpanzee must do with language before its usage qualified as human, but imposing stricter and stricter criteria is little different from arguing that, by definition, only humans are capable of learning to use language as humans do.

A more reasonable approach is to explore with an open mind the possibility of communicating with a nonhuman. It is not unreasonable to assume that some precursor of the ability to use language exists in apes, our nearest biological relatives. In trying to converse with a nonhuman, the worst outcome would be failure, but even failure would force us to rethink our own use of language. In seeking what is distinctive about language, we would acquire a better understanding of our nature, and any difference between our and an ape's abilities to use language could be described objectively, without recourse to a smug sense of superiority. That attitude only serves to mask our ignorance of what we really are.

My plan for teaching Nim to use sign language called for raising him as a human child in a human family. Some attempts to teach language to chimpanzees have lost sight of the importance of establishing language within the context of one or more social bonds. I felt that it would be foolhardy to overlook the obvious fact that a human child learns language as a byproduct of its socialization. I hoped that Nim's motivation to sign would be similar to a child's motivation to talk: not just to communicate his feelings and desires, but to please his family and to share his perceptions of the world.

I envisioned two major differences between Nim's upbringing and that of a normal human child: the people who looked after him would communicate in sign language both with him and among themselves, and his family would be somewhat larger than a normal human family. Nim needed a caretaker during all his waking hours. Each caretaker would not only have to teach him but would also have to provide scientific data about his signing. That tedious and draining work would have to be shared by more than two "parents."

The necessity of performing this research as a group effort on a nonstop basis resulted in problems rarely encountered in a scientific project. For that reason, the story of Project Nim is more than the story of Nim's linguistic achievements, most of which are summarized in scientific articles in the objective and rigorous style appropriate for such publications. But there remains the vital element of Nim's socialization. Because of their subjective nature, important details of Nim's socialization cannot be described properly in the objective terminology of the "method" section of a scientific article. They require an understanding of the human setting of the experiment, of the people who took part in it and the places in which they worked. They also require some understanding of Nim's personality, as elusive and complicated as that of any human child.

As a scientist, I consider these matters important not only because of what they explain about Nim's grasp of sign language but also because they illuminate the nature of the experiment. Aside from reporting how much Nim learned about sign language, this book will describe the relationships that developed between him and the members of his extended human family, the relationships that developed between its members, and the many financial and personnel crises along the way. It will tell how, as soon as each crisis passed, various family members resumed arguing among themselves about when and how Nim should be disciplined, how he should be taught, and who really understood and loved him the most. This book is about Nim, but it is also about a group of people caught up in a fascinating idea.

2

Background

Only fifteen years ago, the idea of a two-way "conversation" with a non-human creature about anything but its most rudimentary needs would have been regarded as science fiction. This state of affairs was changed dramatically by several experiments on language and learning by chimpanzees. Now for the first time there is a firm basis for thinking that a chimpanzee and a human may be able to converse about a broad range of topics. Were it not for these pioneering studies, I would have had neither a basis for starting my own project, nor the courage to do so. Without understanding that background, it would be easy to credit Project Nim with ideas and discoveries that preceded it or to misunderstand why the project proceeded as it did. It is also important to make explicit certain basic assumptions of psychologists who study language in nonhumans.

Why try to communicate with a chimpanzee as opposed to some household pet? One obvious reason is that of all the primates chimpanzees seem most similar to humans in their intelligence, curiosity, and social habits. With a creature as intelligent and as well-socialized as a chimpanzee, Project Nim and other studies have attempted to overcome the limitations of "conversations" with ordinary household pets.

Dogs, cats, horses, and other animals are able to understand many of the sounds spoken by humans. They also produce certain recognizable sounds when they are happy, frightened, playful, and so on. But these exchanges involve nothing more abstract than simple words. Indeed, there is good reason to argue that a cat's meow or a horse's neigh are quite different from such human utterances as "food" or "frightened." When an animal makes a sound, we can usually be sure of two things: it is responding reflexively, and it is reporting some bodily state or emotion. It is doubtful that a cat could learn to suppress its meow when hungry or to change its meow to another sound in order to express its hunger. Likewise, a horse seems able to make one and only one sound when it is frightened. Humans, on the other hand, have no trouble learning whatever arbitrary word their culture uses to express a particular meaning. "Eat," "manger," and "essen" are but a small fraction of the words found in human languages that describe the consumption of food. Besides, words referring to bodily states constitute but a small fraction of the vocabulary of even the most "primitive" societies.

Even if animals' vocalizations and gestures were as voluntary, plentiful, and flexible as human words, the languages of animals and humans would still differ profoundly. In humans the capacity to learn words is secondary to the ability to combine and recombine them to create new meanings. Imagine the difficulty of understanding this information if it were presented one word at a time. Even in simple conversation, our ability to talk about relationships, about past and future events, about abstract ideas, would be reduced drastically if each of our utterances were limited to a single word.

Sentences provide an unlimited opportunity to create new meanings. While the number of words in a given language is finite, the number of possible sentences is infinite. An unabridged dictionary lists all the words of a particular language and their meanings, but no book could possibly list the meanings of all sentences. In conversation, we are constantly creating new sentences in order to communicate aspects of our lives and experiences. In addition, words are usually learned one at a time. Sentences are not. We need no special training to say, "Look out, the red car is skidding toward you!" While the individual words and possibly the phrases of this sentence had to be learned separately, we are able to generate this particular sequence of words instantly to provide a warning about an exceptional situation.

The creative act of generating new meanings through new combinations of words distinguishes a dog's stereotyped food bark from a child's saying, "I like this ice cream cone more than the one I ate yesterday." Both the dog and the child are "talking" about food, but only the child is able to go beyond its immediate circumstances to produce an orderly sequence of words that conveys a specific meaning. This ability is the basis for the claim that learning a language implies learning a set of rules that governs how words may or may not be combined. Linguists call these rules the grammar of a language and use them to describe its structure. Most people learn to talk correctly without any awareness of the grammar needed to generate the sentences they produce. The important thing is to recognize that our ability to create and comprehend sentences presupposes an ability to conform to a grammar.

If a chimpanzee could truly create and comprehend sentences, it is easy to imagine many new and exciting types of communication between humans and chimps. Aside from telling us about basic emotions and feelings, the chimpanzee might also tell us about its memories of pleasant and unpleasant events, its hopes for the future, and its preferences and dislikes. Most important, we might be able to share complex experiences and knowledge with chimpanzees, which would be impossible if communication were limited to a small number of single words. No doubt the feasibility of meaningful communication with a chimpanzee would be questioned and resisted by those committed to preserving the few remaining distinctions between human and nonhuman creatures. But whatever one's point of view, there is little doubt that a chimpanzee who could grasp some of the essentials of sentence construction would be embarking on a path that had previously seemed closed to nonhumans.

The first experiments aimed at teaching language to chimpanzees were failures. In each case, an infant chimpanzee was raised by a human family and exposed to the same type of stimulation given a human infant. The experimenters hoped that the chimpanzee would mimic the human words it heard and eventually learn to use them correctly, to "ape" the language of the family in which it was raised.

In the most successful study of this type, psychologists Keith and Katherine Hayes trained Vicki, their adopted female chimpanzee, to say four words: "momma," "poppa," "up," and "cup." Even these few words were produced in a labored and unnatural manner and required considerable coaxing. In order to get Vicki to use the words, her teachers literally had to form her lips and mouth in the correct articulatory positions. Vicki finally learned to position her lips and mouth with her own hands in order to produce the sounds that her surrogate parents demanded. But even if Vicki had been able to produce these sounds in a humanlike manner, it would have been wrong to conclude that she understood their meanings. Vicki would imitate each of these words only after it was uttered by one of the Hayeses and only if she was given a food or drink reward. What the Hayeses showed was that a chimpanzee could learn some unnatural tricks in order to obtain a reward.

Other studies conducted both in America and the U.S.S.R. were unable to get chimpanzees to say anything. Even when the chimpanzee was raised with a human "sibling" of the same age, and thereby given the opportunity to hear its sibling's baby talk, only the human child learned to speak. These negative results along with Vicki's meager achievements were interpreted widely as evidence that chimpanzees lacked the intelligence to master the abstract complexities of human language. Another possible interpretation, however, was that chimpanzees failed to learn how to talk simply because they could not produce human sounds.

Humans are capable of producing and discriminating approximately one hundred basic sounds. These building blocks of speech are called phonemes. The number of phonemes used by any one language usually ranges between thirty-five and fifty. Chimpanzees can produce only a dozen or so distinct sounds, and these seem specific to particular motivational contexts. Unlike humans, chimpanzees are prone to vocalize only after discovering food, when frightened, during play, while mating, and so on.

The atomic nature of human phonemes is of even greater significance in distinguishing between the vocalization of a human and a chimpanzee, for phonemes can be combined and recombined so as to create thousands of different words. We are able to form words in spoken language by combining sounds in highly systematic ways. A relatively small set of approximately forty sounds is combined in different ways to yield all the words in the English language. Consider the sounds represented by the letters "k," "p," "t," "i," and "a." They can be combined to form the words "pit," "kit," "tip," "pat," "apt," and others. Each combination has a different meaning. There is, however, no evidence that chimpanzees can recombine or reorder the sounds they can produce to form different meanings.

A silent presentation of the Hayeses' film showing Vicki speak her limited vocabulary inspired another husband-wife team of psychologists to use a nonvocal language to communicate with a chimp. Even without the sound track, it was possible to tell what Vicki had said by watching the gestures that accompanied her sounds. Watching Vicki trying to talk and knowing how readily chimpanzees gesture, Allen and Beatrice Gardner, of the University of Nevada in Reno, wondered whether it might be easier to communicate with a chimpanzee through gestures than through spoken sounds.

In June 1966, the Gardners acquired a one-year-old infant female chimpanzee and named her Washoe, after the county in Nevada where they lived. With one major exception, the Gardners provided Washoe with the same type of environment that was provided for the subjects of the earlier experiments. No spoken language was allowed in Washoe's presence. Whatever was communicated to Washoe or between the people who were with her was communicated in American Sign Language. (See Appendix A for a description of American Sign Language.) In addition, Washoe was given the same kind of attention, playthings, games, and tender loving care provided for human infants.

Unlike her predecessors, Washoe did learn to use words of a human language. By the time she left the Gardners, some three and a half years later, Washoe had learned to express 132 words. A list of these signs appears in Table 1. The Gardners estimated that Washoe understood about three times as many signs as she could express. Washoe clearly earned her place in history as the first chimpanzee to communicate with the words of a natural human language.

After learning five signs, Washoe was observed to combine *come gimme, sweet,* and *open* to produce the utterances *Come gimme sweet* and *Come gimme open.* As her vocabulary grew, both the length and the frequency of her combinations increased. Some examples of the many combinations Washoe used are listed in Table 2.

Table 1.
Washoe's Vocabulary

airplane	Don	Larry	same
baby	down	leaf	shoes
banana	Dr. G.	light	smell
bath	drink	Linn	smile
bed	enough	listen	smoke
berry	floor	lock	sorry
bird	flower	lollipop	spin
bite	food-eat	look	spoon
black	fork	man	stamp
blanket	fruit	me	string
book	funny	meat	Susan
brush	go	mine	swallow
bug	good	mirror	sweet
butterfly	goodby	more	telephone
can't	grass	Mrs. G.	there
car	green	Naomi	tickle
cat	Greg	no	time
catch	hammer	nut	tomato
cereal	hand	oil	toothbrush
chair	hat	open	tree
cheese	help	out	up
clean	hole	pants	want
climb	hose	pencil	Washoe
clothes	hot	pin	water
comb	house	pipe	we
come-gimme	hug	please	Wende
cow	hurry	purse	white
cry	hurt	quiet	who
cucumber	ice	red	window
Dennis	in	ride	wiper
different	key	Roger	woman
dirty	kiss	Ron	you
dog	knife	run	yours

Table 2.
Examples of Washoe's Two-Sign Combinations

drink red	go in	tickle Washoe
comb black	look out	open blanket
Washoe sorry	go flower	please tickle
Naomi good	pants tickle	hug hurry
clothes Mrs. G.	baby down	gimme flower
you hat	in hat	more fruit
baby mine	Roger tickle	water bird
clothes yours	you drink	dirty monkey

At first glance, many of them seem quite similar to the first sentences of a child. But whether Washoe's utterances qualify as sentences remains an elusive and tantalizing problem. That question cannot be answered by the Gardners' data because the Gardners made no distinction between combinations containing the same signs in different orders. In most cases the Gardners recorded the signs of a combination in the sequence in which they would have occurred had they been uttered by an English-speaking person. For example, *tickle me* and *me tickle* were considered to be the same combination and were therefore recorded as *tickle me*.

In spoken language, different meanings can be generated by changing word order. Though sign language relies on word order as a grammatical device to a lesser extent than does spoken language, word order still plays an important role. Without it, we would not be able to distinguish between the following grammatical and ungrammatical sentences as expressed in English or in American Sign Language:

John sees Bill.	grammatical
Bill sees John.	grammatical
*John Bill sees.**	ungrammatical
*Bill John sees.**	ungrammatical
Sees Bill John.	ungrammatical
Sees John Bill.	ungrammatical

---

* In certain circumstances, special grammatical devices (such as reference to a spatial framework in front of the signer) allow the recipient of the message to grasp the meaning of such sequences, which otherwise would be ambiguous. See Appendix A for further details.

From the limited information provided by the Gardners, we cannot decide if Washoe was emitting signs appropriate to each situation or combining them, according to grammatical rules, to create specific meanings. Suppose, for example, that Washoe signed *tickle me* when she wanted to be tickled. If she generally arranged a verb and an object in that order when she wanted to be the object of a certain action (for example, *tickle me, hug me, chase Washoe, give me*) and reversed that order when she carried out the action (for example, *me tickle, me hug, Washoe chase, Washoe give*), then it might be argued that Washoe had learned a simple grammatical rule. But without information on sign order and how it was used, we have no way of knowing whether Washoe was really creating specific meanings through the use of grammatical rules or simply combining signs at random.

Especially when trying to interpret "creative" combinations, it is important to have detailed information about all instances in which the signs of that combination occurred. Consider for example Washoe's combination *water bird*, which she signed when asked what a swan was. It is difficult for an English-speaking person not to interpret *water bird* as Washoe's combining an adjective and a noun to create a special meaning. A moment's thought, however, reveals a number of simpler interpretations. Washoe had a long history of being asked *What's that?* in the presence of various objects, including bodies of water and birds. Thus, there is no way to tell whether she was signing about a body of water and a bird or a "bird that inhabits water." Even if Washoe was trying to qualify the meaning of "bird," it is important to know whether she favored constructions of the adjective + noun or noun + adjective type. In the absence of such data, reports that Washoe "created" such combinations as *water bird* are of no more significance than reports that a chimpanzee generated a line from a Shakespearean sonnet by banging on a machine whose keys happen to produce different words. The significance of such anecdotes is determined not by the aptness of a few meaningful combinations but by the relative frequency of such combinations among all the utterances that have been observed.

Yet another problem in interpreting Washoe's combinations lies in the circumstances under which she signed. Most of Washoe's signing was in response to signing initiated by her teachers. While this has not been noted in the published reports of the Gardners, it is apparent in a film of Washoe signing. Many of Washoe's replies to her teachers also included some of their signs. For example:

Susan N: *Who smart?*
Washoe: *Smart Washoe*

Susan N: *Smart*
Washoe: *Smart*

Susan N: *Who stupid?*
Washoe: *Susan stupid*
Susan N: *Who?*
Washoe: *Washoe*

Many of Washoe's combinations may be partial or full imitations of her teachers' signs. Even though Washoe's combinations bore a superficial resemblance to human sentences, I could not rule out simpler explanations, which did not presuppose that a chimpanzee had the ability to create a sentence.

At about the same time that the Gardners taught Washoe to use sign language, David Premack, a psychologist at the University of California at Santa Barbara, invented an artificial language of plastic chips and a radically new method for studying the linguistic potential of a chimpanzee. Neither the form nor the color of Premack's plastic chips provided any clue to their meaning. The chip for *apple* was a blue triangle. The chip *same* was a truncated orange rectangle with corrugated edges on two sides. Premack sought to isolate what he regarded as the important "atomic constituents" of language. Then he devised procedures to teach these features, one at a time, to four caged juvenile chimpanzees. His star pupil was Sarah, who began her language training at the age of four.

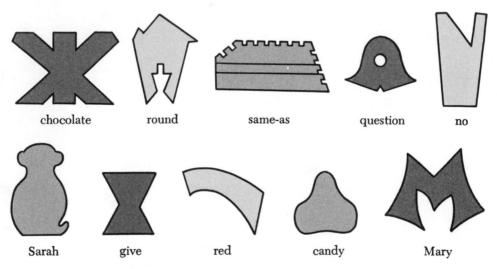

chocolate   round   same-as   question   no

Sarah   give   red   candy   Mary

Some examples of plastic "words" from Sarah's language

Premack's strategy for teaching Sarah was a model of simplicity. Sarah and her teacher sat across from each other, Sarah in her cage and her teacher at the other end of the table separating them. To teach Sarah the name of a food, her teacher would exchange that food for the appropriate plastic chip. When teaching *apple*, Sarah's teacher placed a slice of apple on the table just out of Sarah's reach. After Sarah obtained a few pieces of apple by reaching across the table, her teacher changed the procedure in two ways: a blue plastic chip was placed within easy reach of Sarah, and her teacher would not give her an apple unless she put the blue triangle on the "language board" in front of her. The next step was to have Sarah select the blue triangle from a group of incorrect alternatives that referred to less desirable foods. If Sarah selected a symbol other than the blue triangle, she was given the less desirable food to which that symbol referred. Sarah showed little difficulty in learning to use the blue plastic chip discriminatively in order to obtain pieces of apple.

Once Sarah learned to use a single word, her trainer required that she add additional words before getting a reward. Suppose, for example, that the teacher wanted Sarah to produce the sequence: *Mary give apple Sarah*. First she was required to place *give* next to *apple* on her language board. Then she was required to use these symbols in the correct order. *Give apple* was rewarded, but *apple give* was not. Next Sarah was taught to produce the three-word sequence *Mary give apple*. The two-sign sequence *give apple* was no longer accepted, even when produced in the correct order. To help Sarah learn the names of her trainers, each trainer wore a necklace with a plastic chip referring to his or her name. By seeing which trainer gave her the apple, Sarah was able to match the trainer's name to one of her plastic chips. The last step was to teach Sarah her own name, as in the sequence *Mary give apple Sarah*. By now Sarah was also required to wear her name chip on a necklace. Other chimpanzees were given necklaces symbolizing their names, and the plastic chips referring to their names were also made available to Sarah. One of these chimpanzees, Gussie, was allowed to sit nearby. If Sarah produced the sequence *Mary give apple Gussie*, Gussie got the apple. Sarah rarely made that mistake again.

By means of other simple but ingenious techniques, Premack was able to teach Sarah to use symbols that referred to more abstract concepts—"same," "different," "color of," "name of," and so on. In each case, Premack first made sure that Sarah could differentiate sets of objects that were the same and sets that were different. Sarah's judgment of sameness was checked by giving her a group of objects only two of which were the same, for example, two bowls from the set of objects: bowl 1, bowl 2, and a spoon. She was given a reward if she separated the two identical objects from the one that was different. In the converse problem, Sarah was rewarded for separating the different item or items which were, for example, a sponge from the set of objects: paper clip 1, paper clip 2, and a sponge. In this manner, Premack confirmed that his and Sarah's view of sameness coincided.

When teaching the words *same* and *different*, Premack asked Sarah a very simple question. Two identical objects were placed side by side, separated by a plastic chip meaning ?. The question posed for Sarah was, what is the relationship between the object on the left and the object on the right? All Sarah had to do to answer this question was to replace the ? in the array object 1, ?, object 2, with the only chip available to her. That chip meant *same*. As with all of Sarah's language problems, a correct response was rewarded with food. Questions requiring the response *same* were repeated many times with many pairs of identical objects (two bowls, two paper clips, two matches, and so on). During the next stage of training, Sarah was presented with pairs of different objects. Again she had to answer the question, what is the relationship between these two objects? But this time Sarah was rewarded for inserting the chip meaning *different* between the dissimilar objects. Finally *different* was placed alongside *same*, and Sarah was required to choose between these symbols in reporting whether the objects shown to her were the same or different. Sarah had little difficulty mastering this and related problems.

Premack also taught Sarah to "read" sequences of chips. Each sequence was an instruction about what to do with objects placed in front of her. Sarah might be shown a pail, a dish, an apple, a pear, and the instruction *Sarah insert apple dish*. Sarah was rewarded only if she placed the apple in the dish. She was not rewarded if she placed the apple in the pail or the pear in the pail or the dish. Sarah quickly learned to comply with the following exhaustive set of instructions: *Sarah insert apple dish*; *Sarah insert apple pail*; *Sarah insert pear dish*; *Sarah insert pear pail*. Subsequently she was asked to comply with such compound instructions as *Sarah apple pail pear dish insert*. Here *insert* referred to two actions: the insertion of the apple in the pail and the pear in the dish. Premack regarded Sarah's ability to perform this and related tasks as evidence that Sarah understood the hierarchical relationship between *insert* and the phrases *apple pail* and *pear dish*.

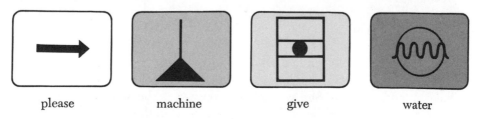

please   machine   give   water

Some Yerkish "words" from Lana's computer console

Lana, a caged chimpanzee, was about two years old when Duane Rumbaugh and some of his colleagues at the Yerkes Primate Center, in Atlanta, Georgia, began to teach her an artificial language that in many ways resembled the language of plastic chips that Premack designed for Sarah. The words of this language, called Yerkish, were available on the keyboard and display console of a computer. When Lana wanted to communicate with the computer, she had to press the appropriate keys on the computer teletypewriter. Each key contained a different lexigram, a unique combination of one of seven colors and one or more of nine geometric configurations. For example, *water* was defined as a red rectangle imprinted with a circle and a wavy line.

At the heart of Rumbaugh's procedure was an infallible computer through which all communication with Lana flowed. The computer always responded instantly to Lana's requests. Because the impassive computer could react in only one manner, it was impossible for the computer to transmit covert "clues" as to the correct response. When human trainers are used, it is difficult to safeguard against transmitting such clues unwittingly. For example, Sarah's trainers might have inadvertently looked at the symbol that was the answer to the problem at hand. By using such clues, Sarah could solve a problem without understanding its meaning.

Lana's lessons on the proper use of Yerkish were quite similar to Sarah's training with plastic chips. If Lana wanted a particular food, she had to ask the computer to provide it by pressing the proper sequence of lexigrams on the computer console. If Lana wanted a piece of apple, she had to press the sequence *please machine give apple*. Lana would not receive any apple if she pressed such incorrect sequences as: *please machine apple give*, or *machine please give apple*. Rewards such as an opportunity to look out the window or to look at pictures of familiar objects required sequences of a slightly different nature. In order to look at slides, Lana would have to type: *please machine show slides*. Sequences such as the following would not be rewarded: *please machine give slides; please machine give window*.

Lana learned not only to "write" correct sequences but also to "read" sequences displayed by the computer. In one test Lana was shown incomplete sequences. If the partial sequence was grammatical (for example, *please machine give* _____), Lana completed the sequence correctly by pressing *apple, Coke,* or *M&M*. She was then given the reward she requested. If the partial sequence was incorrect (for example, *machine please give* _____), Lana gained nothing by continuing the sequence. In order to gain a reward, Lana had to erase the incorrect sequence and begin afresh with a correct sequence. Lana was proficient at either task. She was able to complete correct partial sequences and also to erase incorrect partial sequences, which she then replaced with complete sequences that satisfied the rules of Yerkish.

The achievements of Washoe, Sarah, and Lana are by no means unique. The Gardners have taught American Sign Language to at least four other infant chimpanzees. At the Institute for Primate Studies, in Norman, Oklahoma, Roger Fouts, a student of the Gardners, trained at least five chimpanzees of various ages to use American Sign Language. Premack has taught three other chimpanzees to communicate in his language of plastic chips, and Rumbaugh has trained four other chimpanzees to use Yerkish. As this book was going to press, Francine Patterson, a graduate student at Stanford University, published a paper in which she described her success in teaching American Sign Language to an infant gorilla named Koko.

The evidence is clear that apes have some minimal ability to use and understand words in forms of communication totally foreign to them. This is clearly a significant achievement. Only fifteen years ago, widely recognized psychologists and philosophers (among them Eric Lenneberg, Jacob Bronowski, and Susanne Langer) regarded the ability to use arbitrary words as uniquely human. Whatever the nature of the enormous gap that separates human and ape intelligence, its magnitude has been reduced considerably by demonstrations that chimpanzees can learn extensive vocabularies of arbitrary words.

My initial reaction to the achievements of Washoe, Sarah, Lana, and the other apes who followed their trail was mixed. On the one hand, I felt excited by the possibility of a fundamentally new type of communication with another species. If chimpanzees could use symbols to encode relationships, feelings, and different features of the environment, there was the potential for communicating with them in ways never thought possible. These possibilities led me to take a closer look at the various reports of language in chimpanzees to see just what was being communicated and under what circumstances.

The closer I looked, the more I regarded many reported instances of language as elaborate tricks for obtaining rewards. When Washoe signed *time*, did she do so out of a sense of time, or had she simply learned a gesture to request food, as in the sequence *time eat? Time* was never contrasted with other related signs, such as *now, later, before, soon,* and so on. As far as I could tell, if Washoe had a sense of time, she was not expressing it when she signed *time*. Instead, she seemed to be imitating her teacher, who had just asked her, *time eat?* Adding a meaningless sign to the sign for *eat* is hardly the same as saying, "It is now time to eat."

Similar problems exist in Sarah's and Lana's use of language. It seems doubtful that Lana really understood the meaning of *please* or *machine* in typing the sequence *please machine give apple.* Lana never had the choice to omit *please*, nor was *please* contrasted with other related words. She had as much reason to interpret the lexigram for *please* as "respect" as we have to attribute that meaning to the button on a vending machine. As for Lana's understanding of the lexigram *machine*, she never played with the machine, nor did she have an opportunity to compare it to non-mechanical devices. When Tim, Lana's trainer, was present, Lana received an apple if she performed the sequence *please Tim give apple* rather than *please machine give apple*. That she learned to do so, however, cannot be regarded as evidence that she understood the meaning of *machine* or, for that matter, of Tim's name. Before concluding that *Tim* functioned as Tim's name, one would have to see whether *Tim* could be differentiated from the names of other trainers and whether *Tim* could be used in a wide variety of novel sequences other than *please Tim give X*.

A similar problem of interpretation is encountered when one considers the meaning of Sarah's sequence *Mary give Sarah apple*. It is doubtful that Sarah understood the meanings of *Mary, give,* and *Sarah* in producing this and other sequences of the form *Mary give Sarah X*. *Mary* was simply the first symbol of this rotely learned sequence. In order to select that symbol, all Sarah had to do was to match the symbol that Mary was wearing around her neck with one of the alternatives available on the language board. At the time Premack was training Sarah to use four-symbol sequences, Sarah was unable to differentiate *give* from *insert*, the only other verb she was trained to use. Like *Mary, give* had the status of a nonsense symbol. During training on four-symbol sequences, Sarah was the only chimpanzee present, so it was always correct for Sarah to use the symbol *Sarah* as the third symbol of *Mary give Sarah X*. The only symbol having the status of a meaningful word is the last. In this instance, there is ample evidence that Sarah understood that the last symbol could refer to many different objects.

Creating a sentence presupposes the ability to produce a sequence of words, but all word sequences are not necessarily sentences. They may be nothing more than random combinations or rotely learned chains. It is also foolhardy to conclude that a chimpanzee understands a sentence just because it responds correctly to an instruction presented in the form of a series of words. In each case of purported sentence comprehension on the part of Washoe, Sarah, and Lana, it seems possible that simpler strategies enabled the chimpanzee to solve the problem at hand without understanding all the symbols of the instruction or the relationships between those symbols that an English-speaking observer would normally infer. For example, in the instruction, *Sarah apple pail banana dish insert,* it is unnecessary for Sarah to understand the meaning of either *Sarah* or of *insert*, since there was no other chimpanzee present during the test and the pail and the dish were empty. All Sarah had to understand was that

she had to place certain fruits in certain receptacles. Sarah could determine which fruit went where by observing that the symbol for *apple*
was next to the symbol for *pail* and the symbol for *banana* next to the
symbol for *dish*.

Other examples of Sarah's and Lana's "linguistic behavior" made me
even more doubtful that their various performances required the ability
to create and understand sentences. Sarah's and Lana's production of
sequences like *Mary give Sarah apple* and *please machine give apple*
simply suggested that through rote repetition chimpanzees, like many animals studied in conditioning experiments, were capable of learning a chain
of responses. Suppose that a pigeon rather than a chimpanzee produced
a series of responses. For example, the pigeon might be taught to peck
different colors in a certain order: green → white → red → blue. Having
witnessed such behavior it would be unreasonable to conclude that the
pigeon produced a sentence. Matters wouldn't change even if we super-
imposed the word "please" on the green stimulus, "machine" on the white
stimulus, "give" on the red stimulus, and "food" on the blue stimulus. The
conclusion is still the same. The pigeon learned that in order to get food
it had to peck four colors in a particular sequence.

In collaboration with Rick Straub, Mark Seidenberg, and Tom
Bever, I have recently performed the above experiment. Pigeons were
trained to peck four simultaneously available colors in the sequence
green → white → red → blue, irrespective of their position. This type of
behavior seemed similar to Lana's punching the buttons on the computer
console in the order *please machine give apple* or Sarah's producing the
sequence *Mary give Sarah apple*. While Lana and Sarah each learned to
make a few substitutions for particular incentives, their sequences tend to
read like variations of the kind of rotely learned chains that can be trained
in lower animals. It did not seem to me as if Sarah's and Lana's sequences
display the creativity and broad range of meanings that are so obviously
characteristic of a child's sentences.

The rich linguistic knowledge implied when a child utters a four-
word sentence such as "Mommy give me milk" is apparent when we consider other sentences the child is able to generate: "Sally give me soda,"
"Daddy give cat milk," "I throw dog ball," and so on. In any of the
four positions of the original sentence, the child can substitute a variety
of words, each appropriate to the circumstances at hand. Such sophistica-
tion was not demonstrated by chimpanzees.

In evaluating the evidence that Sarah and Lana learned rudimentary
rules of language, it is important not to lose sight of the narrowness of
their training conditions. During any one session Sarah was given only
one kind of problem (four-symbol sequences, *same* versus *different*, and
so on). Each problem could be answered by choosing from a small set
of alternative symbols. Lana's training regimen was only slightly broader.

Premack's and Rumbaugh's methods are undoubtedly efficient in getting chimpanzees to solve problems. Problem solving, however, is not the same as language.

Another disturbing feature of Sarah's use of language is that it was motivated by food and drink rewards. Lana's motivation for using Yerkish was only slightly broader. She also used Yerkish to have someone groom or tickle her, or to get a chance to look out a window or at pictures. But even with this expanded set of rewards, Lana's and Sarah's use of language can be reduced to requests for specific objects or events. Neither Sarah nor Lana used their languages if they were uninterested in their teachers' stock of rewards. Nor did they comment spontaneously about people or things in their environments. At best, they would solve certain problems if they fancied the rewards they could earn by so doing.

Given Sarah's and Lana's asocial environments, their limited motivation to use language should not seem very surprising. Human language develops out of complex social interactions between a child and its parents. Of course, a child will ask for certain things just as Sarah and Lana did. But a child will also create sentences like "All gone milk," "Bye-bye, Daddy," "Here comes the mailman," "Where's the ball?," "That's a nice doggy," and hundreds of other comments and questions about its environment. In addition, a child uses language to exchange information and to intensify its social bonds with its parents. Indeed, what gives human language its richness and complexity is the extent to which it is used for purposes other than the immediate gratification of basic needs. Were it not for the need to exchange information, it is doubtful that our ability to create sentences would have developed to the extent that it has.

At first glance, Washoe's signing did not seem to be limited to the satisfaction of basic needs. That would seem to follow from the differences between hers and Sarah's and Lana's socialization. When asked to, Washoe named various objects—flowers, dolls, dogs, and so on. However, films of Washoe's signing suggest that those instances had to be prompted by her teachers with insistent requests such as *What's that?* or *Sign!* Also, unless Washoe signed, she would not be given the flower or the doll or be allowed to play with the dog she wanted. In some instances, it seemed that Washoe *had* to be given little pieces of candy before she would sign about the objects her teachers showed her.

Having read the publications of the Gardners, Premack, and Rumbaugh, I was convinced that the state of the art of teaching chimpanzees to use language had advanced considerably by using nonvocal languages compatible with the physical nature of a chimpanzee. There was an abundance of evidence showing that a chimpanzee could learn to acquire a respectable vocabulary of words whose form had no obvious relationship to the objects they symbolized. I remained skeptical, however, about the evidence the Gardners, Premack, and Rumbaugh presented that implied that a chimpanzee could create sentences or that their motivation to use language was sufficient to allow them to engage in conversations about things other than their basic needs. At the same time, the achievements of Washoe, Sarah, and Lana, the high degree of intelligence in chimpanzees, and the degree to which humans seemed to be able to figure out their feelings and moods (and vice versa) gave me hope that chimpanzees could be taught to use language in a humanlike manner. That hope prompted me to start Project Nim.

3

Beginnings

I first met Nim Chimpsky on December 2, 1973, when he was two weeks old. He had just been flown to New York from his birthplace in Norman, Oklahoma, by Stephanie LaFarge, the woman who would serve as his surrogate mother for the next year and a half. When Stephanie first put the furry, cuddly bundle into my arms, I had no idea what a decisive influence Nim would have on the next few years of my life or what a complex and exciting project would develop around him.

Nim was not the first chimpanzee I had adopted, nor was I a stranger to the demands of research. Just the same, I was much less prepared for the demands of Project Nim than I had suspected. In hindsight, it is all too easy to see how severely I had underestimated the difficulties that lay ahead, partly because the circumstances for starting the project seemed quite favorable. Stephanie LaFarge, a highly talented teacher who had raised a chimpanzee for me during an earlier study and who also knew sign language, was eager to look after Nim. Ever since I had come to Columbia in 1961, I had managed to secure continuous and ample funding for my research. During fifteen years of behavioral research, I had also followed with keen interest developments and controversies concerning the nature of language, particularly how it is learned. Tom Bever, a colleague and an outstanding psycholinguist, agreed to collaborate with me in writing a grant proposal. I even had a secretary who was fluent enough in sign language to teach it to me and to any of the volunteers who would join the project.

All of these favorable circumstances were outweighed by Nim's unexpected birth. From my point of view, Nim was an unplanned baby. When he was born, I had barely begun to plan a project centered around an infant chimpanzee that I hoped to acquire during the summer or fall of 1974. All might have gone according to schedule if not for my previous relationship with Nim's half brother Bruno. It was with Bruno, some six years earlier, that I had first explored the reality of rearing an infant chimpanzee in a human setting.

Bruno was born in 1968, about a year and a half after I first heard of Allen and Beatrice Gardner's attempt to break through the language barrier between humans and chimpanzees with sign language. Even before the full scope of the Gardners' project was published, I suspected that their clever strategy would enable a chimpanzee to use language to a much greater extent than previously thought possible. In November 1967, a lecture trip to the West Coast gave me an excuse to ask the Gardners if I could visit them and observe Washoe signing first-hand. They kindly agreed and gave me a crash course on Washoe's signs prior to my meeting her. But knowing Washoe's signs did not prove very useful until she got to know me on a more basic primate level. Upon seeing me, Washoe jumped into my arms and began exploring my face and my clothing. She then urged me to tickle and to chase her.

After a few rounds of play, Washoe led me into the garage of the Gardners' house, used as an informal laboratory where Washoe's teachers taught her to sign and where Allen Gardner photographed her from behind a blind. While I did see Washoe make some signs, my inability to sign fluently and to read her signs without hesitation made it impossible for us to engage in anything approaching a conversation. Nevertheless, when I left Washoe and the Gardners I was more intrigued than ever by the possibility of communicating with a chimpanzee. In order to learn more about chimpanzees, after my return to New York, I arranged to visit the Institute for Primate Studies at Norman, Oklahoma. There, Director William Lemmon graciously spent two days talking about and showing me his collection of more than a dozen chimpanzees.

One of Dr. Lemmon's special interests was the social development and personalities of infant chimpanzees. In one study he compared infant chimpanzees raised by their natural mothers with those raised by human surrogate parents who adopted their foster ape within two weeks of its birth. Dr. Lemmon noticed that chimpanzees raised by human parents developed more rapidly than chimpanzees raised by their natural mothers. Smiling began at an earlier age in home-reared chimpanzees. A home-reared chimpanzee gazed at its surrogate mother's face at an earlier age than did a chimpanzee raised by its natural mother. Dr. Lemmon also observed that home-reared chimpanzees were able to track a light in the dark at a much younger age.

Particularly intriguing was Dr. Lemmon's observation that the personalities of chimpanzees varied as much as the personalities of humans. After meeting three chimpanzees raised by foster parents, I felt that there was a sound basis for Dr. Lemmon's assessment. One of the chimps I met, who was brought up by loving but strict foster parents, was very docile and well behaved. Another, raised by overindulgent foster parents, acted like a spoiled brat. The third was very tense and nervous, particularly when its overanxious foster parents showed signs of leaving the room. Dr. Lemmon also told me how two young female chimpanzees had died suddenly, apparently from depression caused by the sudden substitution of new baby-sitters while the foster parents went away for a weekend.

My visit to Dr. Lemmon ended on a positive note. He told me that if I wanted to try to raise an infant chimpanzee, he would happily furnish one free of charge. The only condition was that I return the animal to the institute when I was no longer able to handle it. This offer gave me the hope of setting up my own project to teach a chimp to use sign language. As a first step, I decided to test the feasibility of raising and socializing an infant chimpanzee in New York. Chimpanzees had a reputation of being especially vulnerable to respiratory ailments. Given the contrast between the cold winters in New York and the year-round warmth of a chimp's natural habitat, the poor quality of New York's atmosphere, and its seasonal rounds of colds and the flu, the question of an infant chimpanzee's survival in that unnatural environment was one I had to answer.

As a bachelor, I obviously could not undertake by myself the task of socializing an infant chimpanzee within a human family. Fortunately, I had a long-standing offer from Stephanie Lee, a former student in a large class in introductory psychology. Stephanie had been the best student in that class and had often asked if I could find some way of including her on a research project. As her neighbor on New York's West Side, I also liked what I had seen of her way with her own three children.

I was confident that Stephanie would make a first-rate surrogate mother for a chimpanzee. No one I knew approached her maternal qualities and enthusiasm for psychological research. I was surprised only by the intensity of Stephanie's "yes" when I asked her if she was interested in raising a chimpanzee with her family. I envisioned a trial period of a little more than a year, during which the infant chimp would be brought up as much as possible like a human infant and thus thoroughly socialized. Socialization, not language, was the point of this project. I made it clear to Stephanie that the program would have to end by June 1969, because I had made firm plans to spend a sabbatical leave in England.

On February 12, 1968, Dr. Lemmon called to ask if I would like to adopt a newborn male chimpanzee. I checked with Stephanie and her family and with two of my graduate students at Columbia who had also expressed an interest in baby-sitting for a newborn chimp. Everyone was still enthusiastic. I told Dr. Lemmon that I would pick up his latest infant sometime in March. During the interim, the baby was separated from his natural mother and cared for by Dr. Lemmon, his wife, and his secretary.

Even before they saw him, Stephanie and her family named the new chimp Bruno. When he was little more than five weeks old, I flew to Oklahoma to bring him to New York and to learn first-hand what it would be like to care for an infant chimpanzee. For the return flight to New York, I reserved a first-class seat in order to minimize the chances of someone making a fuss over him. Acting as if I were carrying a new-born infant, I nonchalantly boarded the plane. Bruno slept in a bassinette on the floor just in front of my seat during most of the flight. When he woke up, ready for a feeding, the stewardess provided a warm bottle without incident. Everything went smoothly until I was getting my things together during the approach to New York. At that time one of the stewardesses came over to "look at my baby" and discovered Bruno. I will never forget the look of sheer horror and the shriek that followed. Within seconds I was reassuring the co-pilot that Bruno, an innocent five weeks old, could cause no harm.

Bruno was a joy to Stephanie, her husband and three children, and Dorrit Lyngsie, a talented *au-pair* girl who moved in after Bruno arrived. Bruno behaved well and was well liked by everyone, but it was with Dorrit and Stephanie's youngest child, Josh, then five years old, that Bruno developed his closest bonds. Josh still remembers his "younger brother." So readily did Bruno and Josh seem to understand each other that I often thought they had developed a special language that only the two of them could understand.

Feeding, burping, and diapering Bruno seemed no different from feeding, burping, and diapering a human infant, an impression confirmed by experienced mothers like Mrs. Lemmon and Stephanie. For at least the first six months, it seemed that human and chimpanzee infants require essentially the same care. Aside from one worrisome bout of pneumonia, I experienced no cause for concern about the health of an infant chimpanzee in New York City.

In June 1969, I returned Bruno to Oklahoma as planned, some four-teen delightful months after he arrived in New York. When it came time for Bruno to leave, we parted company with a healthy and, from our point of view, well-socialized chimpanzee. Since none of us had any experience with other chimpanzees, it was hard to judge just how well we had done with him. But after Bruno reached Oklahoma, Dr. Lemmon told us that Bruno was just about the best-socialized chimp he had ever seen. No doubt this was why he offered to give me another. I thanked him but added that I would have to wait until I could orga-nize a long-term project whose goal would be to teach a chimpanzee to use sign language.

Three years elapsed before I could find the time to do this. During one of those years I was a visiting professor at Harvard University. There, stimulating discussions with B. F. Skinner and Roger Brown about the nature of language gave me special impetus for starting my own project. While a graduate student of Skinner's, twelve years earlier, I had read his classic statement of the behaviorist view of language, *Verbal Behavior*. Skinner's ideas about language hadn't changed much in twelve years. From his point of view, language is not fundamentally different from other types of behavior whose frequency is determined primarily by experience with rewards. A child learns to say "doggy" when that response is rewarded by the parent's saying "good" or by a smile, particularly when a dog is present. Alternatively, the less a child uses language— or the less it engages in any behavior desired by the parent—the less it receives attention and other social rewards.

Deficiencies in the behaviorist approach to language have been noted by many psychologists and philosophers. In 1959 they were reiterated with unusual (if not excessive) vigor by Noam Chomsky in his review of Skinner's book. While I accepted some of Chomsky's criticisms of behaviorism, I felt that his attack was overstated and misdirected. Chomsky's review served as a vehicle for promoting his own theory of language, a theory that says that our ability to create and understand sentences is a direct manifestation of innate mechanisms that are uniquely human. From Chomsky's point of view, only humans have the intellectual capacity to create sentences according to *any* grammar.

For more than a decade, Roger Brown had been studying how a child actually *learns* to use language, a topic that psychologists had neglected, in their rush to create theories of a child's use of language. In 1973 I was able to read a mimeographed draft of Brown's book on the subject, "A First Language." Not only did Brown painstakingly analyze the ways in which children first combine words to create primitive sentences, but he also tried to show how a child's use of language was essentially different from that of a chimpanzee. Though Brown's position was not as extreme as Chomsky's, he doubted that a chimpanzee could create a sentence. But at least Brown's doubt was based upon interpretations of actual data and not on an *a priori* philosophy of human language.

Having absorbed as much as I could from Skinner's and Chomsky's extreme philosophical positions, I left Harvard with a plan to collect the same type of data from a chimpanzee that Brown and other psycholinguists had collected from children. I felt that questions about the linguistic ability of chimpanzees deserved to be answered with facts rather than theories. In September 1973, the start of the academic year following my year at Harvard, I asked Stephanie whether she would like to join me in trying to teach sign language to an infant chimpanzee. Stephanie had been recently re-married to William Ellison Richard LaFarge (who prefers to be called WER) and had moved into a spacious town house on New York's Upper West Side. This time I was not surprised by Stephanie's enthusiastic "yes." Ever since Bruno's return to

Oklahoma, Stephanie had been reminding me that she would like to raise another chimpanzee. Tom Bever, a student of Chomsky's who had joined the Columbia faculty in 1970, also showed interest in studying language acquisition in chimpanzees. Tom's extensive background in linguistics and psycholinguistics contrasted sharply with my background in animal learning. I thought it would be stimulating to collaborate on a project whose subject matter had been a central issue between Skinner and Chomsky, our respective teachers in graduate school.

Dr. Lemmon was not very encouraging when I first asked about the possibility of obtaining another male chimpanzee. (I wanted a male for a number of reasons: I was concerned about Dr. Lemmon's belief that females were more sensitive to depression than males; and since Washoe, Sarah, and Lana, the subjects of the other projects on language in chimpanzees, were all female, I thought it would be of value to investigate linguistic competence in a male.) Dr. Lemmon told me that he had just taken his four females off the pill and that he didn't expect any of them to conceive until November. This meant that no births were expected until May or June of 1974, and he could not promise that any of the babies would be male.

But nine months, the minimum I would have to wait, would not be wasted. In fact, it hardly seemed sufficient to get things ready for a new chimpanzee. At the very least, I wanted to have written a grant proposal for this new line of research. Having written many time-consuming proposals on topics with which I was more familiar, I expected it would take at least two months to write an adequate proposal on a new topic. I would have to summarize what was known about how chimpanzees communicate in their natural habitat, about current and past work on chimpanzee language and intelligence, and about language learning in children. I would also have to provide a detailed plan of study. And once the proposal was finished, I would have to recruit a small staff of assistants and arrange for them to become fluent in sign language.

This orderly program was interrupted by Dr. Lemmon's call announcing the unexpected birth of a male chimpanzee on November 22, 1973. It seemed that the pill was not so reliable a contraceptive for chimpanzee females as it was for human females. Even though I was totally unprepared for a new chimpanzee, Dr. Lemmon urged that I carefully consider adopting him. The father of the newborn chimpanzee was Bruno's father, a male named Pan. Stephanie and I were both impressed by Bruno's intelligence and alertness. The new chimp's mother was Carolyn. Dr. Lemmon regarded Pan and Carolyn as the most intelligent and stable of all the chimps in his colony. Three of Pan and Carolyn's other six offspring had been taught a few dozen signs by Roger Fouts even though they were raised in relative social isolation, in cages or on a man-made island on Dr. Lemmon's farm. In view of Bruno's capacity to learn some thirty words of sign language under nonoptimal conditions, and the similar achievements of Ally, Onan, and Tania, the new chimp's full siblings, it seemed reasonable to expect that Carolyn's newborn chimp would also be quite intelligent. Dr. Lemmon also told me that he could not predict when the next birth in his colony would occur. (His uncertainty was well founded. It was not until May 1974 that the next chimp was born at the institute, and the next male was not born until January 1975.)

My initial inclination was to wait. I could not see how anything could justify jumping ahead with so little planning. But when I told Stephanie and Tom that a newborn male chimpanzee was available, they both argued against waiting. Stephanie, who was not going to school at the time, wanted to start graduate school the following year. From her point of view it made sense to establish a bond with a new chimpanzee while she was relatively free. She also assured me that her family, especially her younger daughter, Jennie, then thirteen years old, would have lots of time to devote to a new chimpanzee.

Tom pointed out that the questions I wanted to study, in particular the question of whether a chimpanzee could create a sentence, might be answered by other projects if I decided to wait. He offered to help by sharing the writing of the first grant proposal. My secretary, Connie Garlock, whose judgment I respected highly, also urged me to go ahead. Connie was a psychology major fluent in sign language. In addition to offering to teach sign language to me and anyone else I was able to interest in the project, she volunteered to work with the new chimp herself.

Despite these assurances, I was still worried. It was not my style to start a new project until I was sure that there would be enough resources to insure that the new project had a good chance of success. I also wanted to acquire sufficient expertise in linguistics and sign language before plunging in.

Still looking for a reason to put things off, I spoke with my departmental chairman, Julian Hochberg, a former teacher and a friend. I asked him for his advice and what I could count on from the department if I needed additional help. Julie was unexpectedly encouraging. He was aware of my long-standing interest in studying language in chimpanzees and thought that I might make a significant contribution. He promised to do whatever he could to help the project, a promise he fulfilled on a number of crucial occasions.

Unable to stall any longer, I agreed to adopt Carolyn's and Pan's newest offspring. I christened him *in absentia* Nim Chimpsky. This name obviously alludes to Noam Chomsky. (Originally, I proposed the names Noam Chimpsky or Neam Chimpsky. But Tom Bever, who had studied under Chomsky, pointed out that from a linguistic point of view, if

Chomsky was changed to Chimpsky, Noam should be changed to Niim or to its shorter equivalent, Nim. As a single name, Nim also had a jungle quality and seemed appropriate even when not accompanied by the surname Chimpsky.)

On November 27, 1973, Stephanie flew to Norman, Oklahoma, to bring Nim to New York. When she first saw Nim he was still with his mother, Carolyn. Having had other babies removed from her, Carolyn seemed apprehensive when Stephanie and Dr. Lemmon approached her. Carolyn must have communicated her fear to the chimps in the neighboring cages, for they soon began to act in an agitated manner and hooted loudly as Dr. Lemmon fired a tranquilizer dart at Carolyn. A short time later Carolyn fell over, briefly asleep, and Dr. Lemmon and Stephanie rushed into the cage to make sure that Nim wouldn't be accidentally crushed by his mother. Stephanie gathered Nim up in a blanket and took him to her motel room, where she diapered him and gave him a bottle. A few days later she and Nim flew to New York. Project Nim had begun.

4

Nim's First Nine Months, His First Six Signs

Because Nim was an unplanned baby, much of his early caretaking had to be improvised. Just the same, I was able to put into effect most of my plan for raising a chimpanzee in a human environment. Thanks to the willingness of Stephanie, her family, and many volunteers to accommodate their lives to Nim's needs, his first year conformed quite closely to the goals of the hastily assembled project that evolved around him.

The goals of Project Nim are easily stated. I wanted to socialize a chimpanzee so that he would be just as concerned about his status in the eyes of his caretakers as he would about the food and drink they had the power to dispense. By making our feelings and reactions a source of concern to Nim, I felt that we could motivate him to use sign language, not just to demand things, but also to describe his feelings and to tell us about his views of people and objects. I wanted to see what combinations of signs Nim would produce without special training, that is, with no more encouragement than the praise that a child receives from its parents. I especially wanted to find out whether these combinations would be similar to human sentences in the sense that they were generated by some grammatical rule.

I also wanted a well-disciplined chimpanzee with whom I could work indefinitely. The longer I could keep Nim, the more opportunities I would have to observe his signing. The importance of being able to work with Nim over a long period of time becomes apparent by considering what an extraterrestrial scientist might conclude about human language under circumstances similar to that of Project Nim. What if the scientist's information consisted entirely of data obtained during the first four years of a child's life? Our hypothetical visitor from outer space would depart with a very limited picture of the human potential for language. Nim's knowledge of language would almost certainly develop more slowly than that of a child. If I could keep Nim for no more than two or three years, I could expect to obtain only a limited picture of his linguistic ability.

The life expectancy of chimpanzees who grow up in their natural environments has been estimated to range from thirty to forty years, a life expectancy similar to that of man at the beginning of his recorded history. Zoos report similar estimates of longevity for chimpanzees raised in captivity. The life of a chimpanzee can be broken down into four distinct phases. During infancy, approximately four years, the chimpanzee is fed and looked after by its mother. It then enters a juvenile phase during which it develops some independence from its mother and begins to look after itself. That phase ends when the chimpanzee becomes sexually mature, usually during its seventh or eighth year. A long adolescence follows puberty. For a male, that period ends when it can establish dominance over some group of males and females. Dominance means first access to food and to females at the height of their estrous cycle, when they are most likely to conceive. The transition from adolescence to adulthood in females is less clearly defined, though in some instances they have been observed to share dominance (as defined by access to food) with males.

Most home-reared chimpanzees have had to be placed in cages well before they become sexually mature. Even before puberty, a chimpanzee's sheer physical strength and its ability to inflict injury through its powerful canine teeth make it difficult to manage. With special care, however, I thought that it might be possible to keep Nim at least through his early adolescence. I knew of one documented case of a chimpanzee who was raised at home until she was eleven. I had also heard of other chimpanzees being kept in the homes of their human caretakers through adulthood. The key seemed to be the consistent application of a set of rules that would insure control and prevent bodily injury.

To maximize the amount of time I could work with Nim, I would have to teach him rules about breaking things, biting, and other unsocial and potentially harmful acts. In order to achieve these goals I planned to have Nim raised under a strict set of rules, but with all the love that would be given to a human infant.

From my experience with Bruno, Nim's half brother, I had learned that it was not very difficult for a human family to raise a well-disciplined infant chimpanzee in a New York apartment. Because the mother of that family was interested in raising a second chimpanzee, I felt confident that, at the very least, we would be off to a good start in rearing a well-socialized animal. But Project Nim called for more than socialization. Our well-socialized chimpanzee also had to learn sign language, his use of sign language had to be documented thoroughly, and his instruction had to be well coordinated. The difference between raising Bruno and Project Nim was the difference between mere baby-sitting and a scientific project aimed at collecting scientifically useful information about a chimpanzee's use of sign language. The latter proved difficult for a number of reasons.

In the first place, I wanted to document Nim's socialization and his use of sign language in a nonlaboratory setting with something that approached the rigor of a laboratory study. Even if Nim had been raised in my home, I would inevitably have encountered situations in which it would be necessary to relax laboratory standards. In someone else's home these problems were even more difficult to control. Even without Nim, Stephanie's was not the typical nuclear family. Her complex household included nine other people: her second husband, WER; three of her children from an earlier marriage, Heather, Jennifer, and Joshua Lee (aged fifteen, fourteen, and eleven); often WER's four children from an earlier marriage, Louisa, Annik, Albert, and Mathilda (aged fifteen, fourteen, eleven, and eight); and Marika Moosbrugger, a twenty-eight-year-old school teacher who was a close friend of the family.

A rare glimpse of the LaFarge household assembled in one place: WER and Stephanie are standing at the back; Jennie is on the far right.

A major problem I encountered in working with Stephanie's family was reconciling various perceptions of the way Nim should be treated and of the kinds of data to be collected. If I had a difference of opinion with a graduate student about what strategy to follow in research in my laboratory at Columbia, it was usually easy to resolve by reasoning together or by citing references in the relevant literature. In the LaFarge house, however, I could only remind Stephanie, her family, and the volunteers who worked there of general guidelines for dealing with Nim. It is one thing to explain to a graduate student why it is important to switch bird number 36 from condition A to condition B for two weeks and another thing to tell a member of someone else's family never to allow Nim to touch the bookshelves. Under these circumstances it should be clear why it was not possible to control Nim's upbringing—a crucial variable in this study—with anything approaching the rigor of the laboratory.

At the start of Project Nim, Stephanie was faced with the challenge of assimilating a new husband, a new group of children, a new townhouse on New York's West Side, and a burning ambition to start graduate school. What better way than to involve her family in Project Nim as much as possible? Nim would be a new creature whom everybody would share. Stephanie told me that at an earlier stage of her life she might have tried to integrate her new family and their new home by having a baby. But now that she had decided against having any more children, she sought to satisfy her family and her research goals by having a chimpanzee. The arrival of a newborn chimpanzee would be like childbirth without the physiological difficulties of childbirth. And since Stephanie would be helped by numerous volunteer assistants, her role as Nim's mother did not include the more tedious aspects of motherhood. She would, however, be able to experience the intimacy of raising a newborn creature in her own home.

Stephanie's hope that Nim would serve as a focus for her newly formed household ran into difficulties on the first day of Project Nim. After bringing Nim from Oklahoma City to New York, she was met at the airport by WER and her daughter Jennie. As soon as they arrived, Stephanie handed Nim to Jenny and greeted WER. Nim responded positively to Jenny from the moment she held him. She in turn regarded him as a delightful new baby brother. On the other hand, Nim's first reaction to WER was not so positive. He resisted being held by WER and did not relax until he was handed back to Stephanie.

There are a number of possible reasons for Nim's initial reaction to WER. It could have been a specific reaction, an expression of general dislike for men, or an expression of a dislike for men who were affectionate toward Stephanie or who otherwise distracted from the attention Stephanie gave Nim. Nim had been looked after by Stephanie for the preceding week, ever since he had been separated from his natural mother. WER may simply have had the misfortune of being the first adult to intrude on Nim's relationship with Stephanie. I later came to believe, however, that Nim's reaction to WER was more specific than a general concern about the disruption of his relationship with Stephanie. Often Nim's reaction to WER seemed downright Oedipal.

Nim's new home was a four-story brownstone building on a quiet street just west of the Museum of Natural History. The LaFarge house was one of a whole block of similar brownstone houses, each with a front "stoop" of stone steps and a small backyard. Each of the three floors of the LaFarges' brownstone had a similar layout: a front room overlooking the street and a rear room overlooking the yard and the backs of the row of brownstones on the next street.

Nim was originally confined to rooms on the first two floors: the living room, the dining room, and an adjacent kitchen on the parlor floor, and Stephanie's and WER's bedroom on the second floor. During his first few months in the LaFarge house, Nim slept in a crib at the foot of Stephanie's and WER's bed. He slept between fifteen and eighteen hours a day, in three- to four-hour stretches. While he was awake he was pre-occupied mainly with being fed. At first his diet was a standard baby formula. When he was two months old he began to eat semisolid baby food. Indeed, there were more similarities than differences between his existence and that of a newborn human infant. Nim had his diapers changed every few hours, was burped after feeding, tossed in the air, hugged, and held protectively by each of his caretakers.

At first Nim was looked after mainly by Stephanie, Jennie, and Marika. WER also helped. For the most part WER had a positive and a friendly attitude toward Nim. At other times, however, WER simply resented the small but numerous changes in his day-to-day existence that were caused by Nim's presence. He gradually came to regard Nim as an intruder. Nim must have sensed WER's negative feelings, for more and more he resisted being held by WER; at times he even balked at WER's feeding him.

Soon Nim's response to WER became more overt. On one occasion Stephanie and WER were taking a late-afternoon nap on a large water-bed between their living and dining rooms. Nim, who was then about six months old, was lying between them. Both Stephanie and WER thought that Nim was asleep, but when WER reached over to put his arm around Stephanie, Nim instantly stood up and bit WER. On other occasions Nim hooted aggressively at WER, his mouth drawn back with his hair erect, as if he were threatening to bite WER, and his hands flailing in vain attempts to strike him. Such full displays of aggression were all the more remarkable because Nim had never seen or heard the vocal and emotional reactions of an angry chimpanzee.

The author with his new charge at Columbia

Until a chimpanzee can be psychoanalyzed, one can only speculate as to whether an infant male chimpanzee, normally reared only by its mother, is capable of an Oedipal reaction. At the same time it is hard to ignore various compelling observations reported by Jane Goodall and others of infant male chimps attacking their fathers when the fathers copulated with their mothers. And it is hard to ignore the strength and the immediacy of Nim's reaction, repeated on other occasions, to a mild display of affection by his surrogate father for his surrogate mother.

Stephanie took charge of teaching everyone in her household the first signs we wanted Nim to learn, as well as a basic set of signs conducive to human communication. At the same time, I organized a group of student volunteers from Columbia and Barnard Colleges who were required to learn sign language and to study the background of Project Nim. Recruiting volunteers was surprisingly easy. When it became known that I was looking for students to help teach sign language to an infant chimpanzee, far more students volunteered to work on the project than I actually needed. After a series of interviews designed to discourage all but the most qualified and motivated students, I selected six volunteers to help Stephanie look after Nim.

I visited Nim as often as I could, usually four or five times a week, much more often than I had visited Bruno. I felt more concerned about

Nim than I did about Bruno, partly because I planned to keep him longer. I also found Nim very appealing. For a while I thought that my feelings toward Nim were colored by my desire to have a child of my own. In many ways, however, Nim was even more childlike than a child. His reactions, particularly his emotional responses, were stronger than those of a child. When he cried, his screams seemed more powerful. When he clung, his grip seemed more tenacious. And when his attention was caught by some object, his effort to grab that object seemed more determined. With Nim I felt a directness of expression I have often felt from other animals. But unlike the direct and immediate reactions of pet dogs and cats, Nim was curious and expressive in an almost human way. That combination of intensity and intelligence was hard to resist.

I was not alone in my feelings. Stephanie, who had three children of her own, spoke of the strong relationship that developed between herself and Nim. Many of the inhibitions about sharing pleasure with a human

The infant Nim, with Stephanie, and with his own thoughts

baby were reduced with the infant chimpanzee because body contact was stronger. At the same time, there was a distance based on the awareness of the difference between a chimpanzee and a human infant. Stephanie also felt that her reactions to Nim were freer because she was less concerned about whether Nim would like her.

The intensity and directness of Nim's reaction to his caretakers was especially apparent when Nim saw that they were unhappy. From almost the moment he could crawl he would scamper over to a member of the project as soon as he noticed a downcast look. He seemed to vacillate between wanting to be assured that nothing bad was going to happen to *him* and trying to comfort whoever was upset. Tears brought out especially tender behavior on Nim's part. I once saw him rush over to Jennie while she was crying, leap into her arms, and stare intently at her eyes. He then touched her cheeks very gingerly and gently tried to wipe away her tears.

A few days before Nim's first birthday, Stephanie came home very depressed. She had just been to the hospital to visit her father, who was dying of cancer. Stephanie told me that on that occasion, and also after her father died, she felt that Nim's reaction to her grief was more direct and comforting than that of any of the other inhabitants of her household. During this period especially, Nim clung to Stephanie and acted as if he were trying to protect her by pushing away anyone who approached.

Nim's education in sign language began on his first day in Stephanie's home. Our first goal was to teach Nim to understand our signs about events that were important to him. Before Nim was given a bottle, his caretakers signed *drink?* (the thumb, extending from the fist, is touched to the mouth). Before Nim was picked up, his caretakers signed *up?* (the index finger, extended from the outstretched hand, points upward). And when, after his first month, Nim began to gaze at objects of interest and tried to reach for them, his caretaker would sign *give?* (a beckoning motion with the fingers making an arc toward the signer).

Since none of the volunteers nor anyone at Stephanie's house was a fluent signer, Nim was also exposed to a lot of spoken language. Yet he showed little interest in listening to people's conversations. The first few words would often startle him. Indeed, an effective means of getting Nim to stop what he was doing was to shout "no!" or "stop!" Spoken language was a good device for capturing Nim's attention, but the effect only lasted for a few seconds. This was not the case with sign language, particularly if the signers were fluent. I have often seen Nim transfixed, virtually spellbound for long periods of time (up to fifteen minutes), by the fluid motions of skilled signers.

From the time Nim was two months old, his caretakers began to teach him to sign by physically molding his hands. The technique of molding was first used with chimps by the Gardners and by Fouts. As the term implies, the teacher gently takes the chimpanzee's hands and molds them into the configuration of a particular sign. At first Nim resisted having his hands molded, but once he perceived that his teacher wasn't going to harm him, he soon began to relax his hands and show an interest in learning how to sign.

By the time Nim was four months old, Stephanie, Marika, a few volunteers, and I were regularly molding Nim's hands to make the signs *drink, up, sweet, give,* and *more.* Each time Nim was shown his bottle, his hand was molded into the correct configuration for the *drink* sign. Likewise each time Nim wanted to be picked up, his hand was molded into the correct configuration for the *up* sign. The sign for *sweet* was molded initially for sweet baby foods and subsequently for candies and the addition of honey to Nim's cereal. *Give* was molded in situations in which Nim wanted an object such as a book, ball, doll, or some other toy. *More* was molded whenever Nim wanted an activity to reoccur, such as eating, drinking, being tickled, or having a pillow fight.

On February 4, 1974, when Nim was only two and a half months old, Stephanie reported that he had made a spontaneous *drink* sign: when shown his bottle, he signed *drink* without any molding or prompting on Stephanie's part. Stephanie's reaction to Nim's first sign was one of utter shock. She had no idea just when Nim would begin to sign, but she least expected him to start at so tender an age. Before she could recover from the shock of seeing Nim's first *drink* sign, he repeated the sign a few more times.

As convinced as Stephanie was that Nim had signed *drink,* I required stronger evidence that he had learned a sign. It was not that I questioned Stephanie's powers of observation; I simply wanted to be absolutely sure that anyone who reported a sign had not imagined it and had definitely observed something more than an accidental hand movement. Moreover, I wanted to see Nim use his new sign regularly and appropriately.

By the end of February 1974, Nim was making many spontaneous signs. But before I would say that Nim had actually learned a sign, I required the following two criteria to be met: on different occasions, three independent observers had to report its spontaneous occurrence; and the sign had then to occur spontaneously and appropriately on five successive days. According to these rather stringent requirements, Nim mastered his first sign (the sign for *drink*) on March 4, 1974, before he was four months old. During the next six weeks he learned, according to the same criteria, to sign *up, sweet, give,* and *more,* the remaining signs on our initial target list.

Nim's early mastery of signs was very encouraging, particularly since this was my first attempt to teach a chimpanzee to use sign language. But even though Nim learned his first sign before most children speak their first words, I did not regard Nim as a prodigy. There are a number of well-documented reports of deaf children learning their first sign by the age of four months. A few months after Nim began to sign, the Gardners reported that Moja, one of the new chimpanzees they had raised from birth, learned her first sign at the age of four months. There appears to be a simple explanation for the fact that deaf children and chimpanzees learn to sign about a year before most children learn to speak. The development of the coordination needed to control the relatively gross motor movements used in signing is considerably more rapid than the development of the coordination of the many subtle movements of the mouth, lips, and tongue that are needed to produce sounds.

Sign language was but one facet of Nim's early mental development. In his third month he began to gaze at pictures in magazines and books. At that time he was interested mainly in pictures of people, animals, and food. By his fourth month he was coordinated enough to touch, with his index finger, parts of pictures that interested him. On other occasions Nim would place his mouth on a picture as if kissing it. When he touched or mouthed the picture he usually made contact with the critical part of the picture itself. This was true whether the picture was in color or in black and white, big or small, or printed on the top, middle, side, or bottom of the page.

Stephanie and I often gave Nim informal tests to chart other aspects of his mental development. During his third month I would often open and close my mouth or stick out my tongue at Nim. Nim would gaze intently at my face and imitate both of these actions. By the time Nim was four months old he readily followed a moving finger or a favorite toy that I moved back and forth in front of him. He did not, however, seem able to follow a ball that I dropped in front of him. He would first look at my empty hand, which had just held the ball, and then, a second or so later, lower his head to look at the ball that I had caught with my other hand.

Nim's ability to move around developed simultaneously with his signing. By the end of his second month he was crawling both inside and outside of his playpen. During the next month he could stand if he had something to hold on to. By the end of his fourth month he could crawl or walk to any part of Stephanie's house that interested him. He was especially good at chasing Trudge, the very tolerant German shepherd who lived in Stephanie's home. During his fifth month Nim became too mobile to stay in his crib in Stephanie's and WER's bedroom.

In order to provide him with comfortable and secure sleeping quarters, we suspended a hammock from three intersecting poles. The tripod supporting the hammock was surrounded by netting that could be zippered shut when Nim was put to sleep. Since he could not open the zipper and get out of his enclosure, he no longer needed to sleep under Stephanie's watchful eye. His new sleeping quarters were set up in a corner of the dining room, a room in which he felt quite at home. For the first time, Nim was left by himself.

At first he screamed and whimpered for a few minutes after Stephanie or another member of her household put him to sleep. We soon learned, however, that leaving Nim with a bottle or a pacifier was all that was needed to make him feel comfortable enough to go to sleep with no one else around.

By May of 1974, when Nim was barely six months old, I was more than pleased with the way he was integrated into Stephanie's home and the rate at which he was learning to cope with his world. Needless to say, I was especially pleased by the way he was learning to sign and the interest he had shown in activities that seemed conducive to further intellectual growth. There remained, however, many uncertainties about the project's future. Since I had yet to learn the fate of the grant application that Tom Bever and I had submitted at the beginning of March, it was impossible to contemplate hiring skilled personnel to assist me. The current group of volunteers was busy preparing for final exams and making plans for the summer. They needed paying jobs, which I could not offer. Soon they would be going their separate ways without having been replaced. Stephanie, her family, Marika, and I had to work out a plan for looking after Nim and teaching him to sign during the upcoming summer.

Compounding this problem was the meager amount of time that teaching and administering the project left me for working with Nim. I also needed more time not just for Nim but for Stephanie and her family. For some time I had been troubled by the slow deterioration of several important kinds of coordination between Stephanie, her family, and me. These had to do with rearing practices, teaching techniques, and methods of data collection. There had developed a growing divergence between Stephanie's and my views as to what work should have the highest priority.

In hindsight, these differences are easy to understand. Like many other graduate students, myself included, Stephanie had changed her mind a number of times about the area in which she would specialize. Unfortunately for me, Stephanie's change of interest occurred during the year before she was to start graduate school, the first year of Project Nim. When I first spoke with Stephanie about the possibility of her raising a second chimpanzee, she was planning to enroll in the developmental psychology program at Teachers College of Columbia University to work on her Ph.D. under the tutelage of Lois Bloom, a highly regarded psycholinguist known for her careful studies of the emergence of language in children.

While it is unusual for a student to decide on a dissertation topic before starting doctoral studies, it is also unusual for a predoctoral student to have an opportunity to raise an infant chimpanzee. Before I agreed to adopt Nim, Stephanie had told me that she would be thrilled to do a dissertation on the acquisition of sign language in an infant chimpanzee. From my point of view, that motivation was ideal. The standards of research required for a doctoral dissertation, particularly research done in conjunction with someone as painstakingly careful as Lois Bloom, were consistent with the standards I wanted to set for this project.

Within a few months of becoming Nim's surrogate mother, Stephanie's interests shifted away from the psycholinguistic questions that interested me to questions that had more to do with precursors of language. Stephanie wanted to study just how signing emerged from the complex relationships that developed between Nim and herself and other caretakers. While I placed the highest priority on obtaining detailed and objective data about the structure and context of Nim's utterances, Stephanie was becoming more concerned about gathering information on the nature of Nim's reactions to his teachers.

In an oversimplified way, the issue between Stephanie and me might be characterized as the issue between "hard" and "soft" psychology. I was eager to define the scientific objectives of the project, not only because I was in the midst of writing about those objectives in grant proposals, but also because I felt that early decisions about Nim's upbringing and training would constrain what he would and would not do as he grew older. Stephanie on the other hand had a less focused point of view. She was more interested in adopting an observe, wait, and see attitude. Before trying to teach a particular sign or decide on a particular training method, she wanted to videotape Nim's interactions with various teachers in various situations. Only after analyzing these interactions would Stephanie feel prepared to teach Nim to sign.

As a behavioral scientist, I recognized that Stephanie had raised some valid and interesting questions, but I felt that it would be impractical to broaden the scope of the project as she suggested. At the time I could only borrow, beg, or steal videotape equipment. In general I felt that we didn't have the resources to carry out Stephanie's proposals and that we could not afford to lose valuable time marshaling what resources we had. It was my view that those resources should be used to pursue the central question of the project: What can a chimpanzee learn about language?

In May 1974, Stephanie told me that she did not want to assume both the role of Nim's mother and that of his teacher, the two roles I had envisioned for her at the start of the project. Since a parent normally integrates both roles and since Stephanie was an experienced teacher (she was then the principal of a Montessori school), I took for granted that she would take to being a mother and a teacher equally well.

Stephanie's decision had two immediate ramifications. One was that I could not count on the most important person in Nim's life to be his best and most consistent teacher of signs. The other was a series of debates that our different points of view precipitated at the weekly project meetings I held at my apartment. At these meetings we discussed what signs to teach Nim, how the signs were to be taught, what other aspects of his development would be observed and documented, what records should be kept, how Nim would be disciplined, who would be with him and when, prospects for funding, and so on. Even though I did not always agree with Stephanie, I felt that her ideas should get a full airing. As it turned out, the other members of the project agreed with me that to delay working out procedures for teaching Nim to sign would mean losing valuable time that might in the long run limit his linguistic development. But it was not until the end of the school year that we reached that decision, and the volunteers who supported my view would not return until after the summer.

Two pressing problems had to be solved in order for the project to survive until classes resumed in the fall. One was the logistic problem of who would look after Nim once the volunteers left for the summer. The other was to iron out my differences with Stephanie about the best direction for Project Nim to follow.

With these problems in mind, I rented a house in East Hampton, Long Island, large enough for Nim, Stephanie, her family, and me, where I hoped we could work with Nim jointly. At least we would have an immediate basis for resolving our differences. I was to keep Nim in Easthampton until August, where he would be looked after by me, Stephanie, her family, and two new volunteers (Maggie Jakobson and Penny Franklin), who were also spending their summer in East Hampton. In August, Nim would be moved to WER's farm in Rhode Island, where all of Stephanie's and WER's children would gather for their vacation. Stephanie seemed to welcome the opportunity to join me in working with Nim without the distractions of New York. No doubt the thought of a summer house within easy reach of the city and the relaxing beaches of East Hampton were also attractive.

I rented a secluded turn-of-the-century cottage with ample room both inside and out. The spacious grounds were a welcome improvement over the LaFarges' backyard. Adjacent to the property was a large potato

farm, and beyond, one could see the glistening waters of the ocean. Nim
had just begun to feel comfortable running about by himself. As long as
I or another member of the project was in view, he seemed content to
scurry around exploring the spacious lawn that surrounded the house.

During his summer in East Hampton, Nim discovered nature. While
he wandered around outside the house, he often paused to smell and
touch flowers. He also discovered birds. For many minutes at a time, Nim
would gaze up into the branches of a tree as if trying to figure out what
produced the bird songs he heard. If a bird moved away from the tree,
Nim would try to catch it while hooting attack cries at the top of his
lungs. He also tried—and failed—to catch rabbits. Such things occurred
almost every time Nim went outside.

I was provided with occasional relief from looking after Nim when he
stayed with Penny Franklin or Maggie Jakobson, often for days at a time.
Maggie spent many hours trying to interest Nim in her pool but succeeded only in getting him to put one foot in, in a gingerly fashion. He
showed the same aversion to water at the beach. But there Nim at least
liked to play in the sand and chase whatever dogs came by to get a
closer look.

Nim's summer in East Hampton was a time of much play and little
work. For a variety of reasons, Nim made little progress in learning sign
language. For one brief period, Stephanie and I did manage to work
together in teaching Nim to sign. On that occasion, we sat down next to
Nim's eating table, and Stephanie held a jar of baby food while I molded
Nim's hands to form the sign *eat* (the thumb touches the middle two
fingers and the hand is brought to the mouth in that configuration; see
Appendix C for further details). I then gave him a spoon of food. After
a few repetitions, Nim was signing *eat* spontaneously.

Eat was the first sign I taught Nim myself. It was also the only sign
Nim learned that summer. Without the help of other project members, it
was hard to establish the consistency of teaching that I had hoped for
and that was necessary if we were to satisfy the criteria that allowed us
to say that Nim had learned a new sign. Circumstances neither of us
could control prevented Stephanie from spending much time with Nim;
her daughter Jennie had to be hospitalized for six weeks. It was not until
mid-July that Stephanie and Jennie were able to spend much time in
East Hampton. Even then Jennie required a good deal of attention from
her mother. An additional blow for me was the sudden death of an old
friend, Jacob Bronowski, with whom I had had many discussions about
the possibility of teaching language to a chimpanzee. Toward the end
of what was his first vacation in many years, and two days before he was
to visit and observe Nim, Bronowski died of a heart attack.

It was now August and time for Stephanie, WER, Jennie, and Nim
to leave for the LaFarge farm in Rhode Island. My hopes for establishing

Summer in East Hampton: left, with Stephanie; right, with Maggie Jakobson

an effective coordination between Stephanie's and my efforts with Nim and for persuading Stephanie to serve as Nim's main teacher never materialized. Right after Labor Day, almost immediately upon returning from Rhode Island, Stephanie was to start graduate school. This would leave her with even less time than before the summer to work on the project. She wanted to limit her role on the project to looking after Nim's general care in her home and some occasional teaching of signs. The planning and overseeing of Nim's sign education would have to be done by someone else.

It didn't take me long to conclude that I couldn't be that someone else. I was obligated to teach my regular courses to Columbia students, assume administrative duties in the department, and look after my conditioning laboratory, where, for the last twelve years, I had been studying the conditioned behavior of pigeons. It was difficult to turn away from this demanding research for a number of reasons. I had a strong personal interest in completing various ongoing experiments and I felt an obligation to postdoctoral and graduate students to supervise work they had started with me. Even more urgent than the demands of my conditioning laboratory was the problem of raising enough money to insure the continuation of Project Nim. Under these conditions I had to limit my time with Nim to a few hours a week.

What was needed as soon as possible was another person who would supervise Nim's instruction in sign language and the documentation of what he learned. I also had to find space outside of Stephanie's home where a new teacher could work with Nim. Introducing yet another person into Stephanie's townhouse was out of the question.

With no funds, no candidates for a job born abruptly of necessity, and no place for the recipient of that job to work, it was not obvious how the project would proceed once Nim returned to New York after Labor Day. This was the first of many financial and personnel crises that were resolved just when the outlook seemed bleakest. From my department chairman I learned of money against which I could borrow until the project was funded. He also made available a suite of rooms at Columbia that the new sign teacher could use as a teaching area. Finding a teacher took a little longer. Fluent signers command higher salaries, as interpreters or teachers of sign language, than I could offer. I also discovered that many people concerned with the deaf feel a strong antipathy toward research on sign language in chimpanzees because they believe that it implies an insulting parallel between chimpanzees and deaf people. Fortunately, the New York University Deafness Research Center, one of the finest centers for sign language study in the United States, put me in touch with a teacher who had lots of experience teaching sign language and was looking for a job in New York.

When I interviewed Carol Stewart for the job of Nim's main sign teacher, she impressed me very favorably. She had many ideas about teaching signs and was eager to assume that responsibility on Project Nim. Carol was a teacher at the Southbury Training School, in western Connecticut, which specialized in teaching sign language to retarded children. The teaching method Carol followed was based largely on principles of behavior modification. The teacher broke down the desired behavior into small and observable steps, and then proceeded to teach one step at a time by rewarding only correct behavior. I felt that Nim might benefit from Carol's suggestions for applying the techniques of behavior modification, particularly those emphasizing consistency of instruction. That, plus Carol's obvious sincerity and conscientiousness, made her an attractive candidate. I hired her.

Carol's initial task was to develop a curriculum for Nim. Ten other teachers would assist her, seven from our original group. Totally new to the project were Bill Tynan and Kela Stevens, both undergraduates majoring in psychology at Columbia. Bill and Kela had known Carol previously through their work with deaf people. Bill had videotaped children at the Southbury school, and Kela had worked as an interpreter of sign language. Rounding out our group was Laura Petitto, who had begun to work as a volunteer the previous May, just before Nim left for the summer. A senior at Ramapo College in New Jersey, Laura commuted to New York three or four times a week in order to work with Nim.

Though Nim was not aware of them, preparations were being made for extensive changes in his life. When he returned to New York after his leisurely vacation, at the age of ten months Nim would be the only student of a full-time sign teacher who would instruct him in a special new nursery school at Columbia University.

5

Nim's Nursery School

Most children learn to speak before they go to school. Except for the retarded, autistic, or disabled child, learning to speak is a universal consequence of growing up in a speaking household. Learning sign language is no exception. Deaf children raised by deaf parents learn to sign fluently before they have any schooling. Perhaps the best way for an infant chimpanzee to learn sign language would be from its natural mother. But until that is possible the most promising alternative appears to be instruction by surrogate human parents.

Originally I planned to have Nim learn sign language in Stephanie's home as part of his socialization. The responsibility for teaching Nim to sign was to be shared by members of his extended family: Stephanie, her household, the volunteers, and me. I was, of course, aware that Nim would need much more explicit instruction than would the typical child learning to speak or to sign. I assumed that Stephanie, the person most responsible for Nim's day-to-day care, would also take charge of teaching him to sign. I also assumed that all of Nim's sign instruction would take place in his home. By and large, I thought that what works for children in the way of language instruction would also work for Nim.

Now that Stephanie had decided not to continue in her role as Nim's main teacher, two important changes in the conduct of the project had to be implemented as soon as possible after Labor Day. Nim had to get to know Carol and his other new teachers, and I had to prepare Nim's nursery school.

The suite of four rooms I planned to use as both a nursery school and a day-care center required thorough chimp-proofing. False walls had to be constructed to cover pipes and other protuberances on which Nim might climb. Carpets had to be put down to protect Nim from the cold and the hardness of the concrete floors. Having no funds for these renovations, we had to make do with old carpeting that the university was about to discard and the free labor of the volunteers.

Some renovations continued even after Nim began to use the classroom. We learned that door locks had to be chimp-proofed too. First, we tried a chain-lock in combination with a dead-bolt lock. A month later, when Nim succeeded in opening these locks, we installed spring-loaded hooks. Opening the hooks required considerable dexterity, but even they were not completely effective. Nim occasionally managed to slip out and had us chasing him through the halls and stairwells of the psychology department.

The room used as Nim's classroom was bare and small, a mere eight feet square. This was by design. I felt that Nim would not romp around too much in a small area and would be more likely to concentrate on the activities introduced by his teachers. I also felt that a bare room would minimize distractions. One wall contained a large one-way window that allowed observers in an adjacent office to watch Nim without being seen. That one-way window proved invaluable. Without it, I would not have been able to observe directly what Nim's teachers were doing in the classroom, nor would I have been able to train new teachers by showing them first-hand what it was like to work with Nim. Beneath the one-way window was a portal that could house the various cameras I used to provide a permanent record of Nim's signing.

The remaining two rooms of the suite were on the other side of an internal corridor directly across from the classroom and the observation room. One was used as a play area, the other as a storeroom for toys and educational materials. Originally the internal corridor was used as a diaper-changing area. Once Nim was toilet trained (some ten months later), a potty was placed in the corridor just outside the door to the classroom.

Before Nim was brought to his new classroom, he was introduced to Carol, Bill, and Kela, the new members of the project, at Stephanie's house, and he was reintroduced to the volunteers who had worked with him before the summer. Nim was much more active around the LaFarge home after his summer away from New York. Strict rules had to be laid down: he was not permitted to bang on loudspeakers, climb on book-shelves, pull at plants, climb on kitchen counters, and so on. At first, Carol and the other new teachers had their work cut out for them making sure that Nim obeyed these rules. Once Nim perceived that his new teachers meant business, he came to accept and respect them as he did the teachers with whom he was more familiar.

On September 26, 1974, Carol drove Nim to Columbia for his first visit to the classroom. This was not Nim's first visit to Columbia. During the previous spring Stephanie had occasionally brought him to my office for short visits, during which Nim did not seem very aware of his sur-roundings. Most of the time he was content to sit in my lap and play with whatever he found on my desk. But on this occasion he reacted strongly to his new environment. He clung tightly to Carol as she walked around the freshly painted suite of rooms as though he was disoriented. Unlike my office, where he could recognize me and various objects such as plants, shelves, chairs, tables, and books with which he was generally familiar, Nim's nursery school contained nothing familiar. The rooms were small and without windows or furniture. The walls, which were made of concrete cinder blocks, were completely bare. Even after a large breakfast Nim seemed wary of his new surroundings. After a few tentative steps away from Carol, he curled up in her lap and went to sleep.

Within a few days Nim seemed more at ease. For the first time he allowed himself to be separated from his teachers, and he began to show interest in the playthings they brought to their sessions. But he was still not fully comfortable in the classroom. The slightest noise caused him to hoot and to leap into the arms of his teacher. At times Nim was so scared that he tried to hide under his teacher's skirt. Unfortunately, sounds around the classroom were quite plentiful and difficult to control. Behind one wall was the rat room of the psychology department's vivarium. The rats often squealed when they were taken out of their cages. Almost every day, the rats' high-pitched squeals evoked momentary panic in Nim. He would stare in the direction of the sound as though trying to figure out what could have produced it, or he would rock back and forth on the floor. When he stopped rocking, he would frequently have lost interest in what he had been doing before he was startled. Lesser disturbances were caused by sounds from a stairwell adjacent to another wall of the classroom and from the outside corridor. If there had been money for renovations, it would have been a simple matter to soundproof the classroom. As it was, we had to wait until Nim adapted to these noises. That took quite a few months. We were, however, able to salvage something from the disturbances. They were exploited by Nim's teachers in getting Nim to learn *listen*.

During one series of weekly meetings, Carol described what came to be known as her "method" for working with Nim. Her proposal was not restricted to teaching signs but encompassed all aspects of Nim's behavior. Her purpose was to teach Nim what kinds of behavior were expected of him in all situations and to reinforce those behaviors as strongly as possible. Carol was as concerned about minimizing undesirable behavior as she was about establishing desirable behavior. At the age of one year, she felt that Nim was quite malleable. As he grew older, however, it would become more difficult to teach him what was socially acceptable —playing with toys, looking at books, wearing clothes, romping around in his gym—and what was socially unacceptable—breaking things, biting people, screaming too loudly.

In order to emphasize consistency, Carol worked out a number of daily schedules that specified what was expected of Nim in different situations. When he was brought into the classroom, he was expected to hang his hat, sweater, and coat on a hook on the wall near the door, three feet above the floor. At first he was helped off with his outer garments; later he was expected to remove them by himself. Breakfast was one of the first activities. At breakfast, as well as at other meals, Nim was expected to sit in a high chair, to eat with a spoon, and to wipe both his face and his high chair when finished. After breakfast the teacher would show him various picture books, toys, and other objects in order to see what interested him. The teacher tried to encourage Nim to hold the object he was looking at by himself. If he selected a book, he was encouraged to turn the pages, but he was discouraged from biting the book. In each instance Carol sought to establish socially acceptable behavior in situations conducive to signing.

Carol suggested that sign language be introduced to Nim in three stages. During "reception," the first stage, the teacher tried to get Nim to understand the sign, which meant that he had to make a response that the teacher considered appropriate. If the teacher signed *eat*, Nim was supposed to pick up a piece of food that the teacher provided, or he might be required to select a picture of some food item from a group of pictures that also included pictures of nonfood items.

The purpose of "production," the second stage, was to show Nim how to make a well-formed sign outside the context in which that sign was to be used. First Nim's hands were molded by the teacher to make the sign. Later he was required to make the sign on his own. Carol felt

With Carol

that if Nim was taught the hand configurations needed for different signs out of context, he would learn the mechanics of signing without being excited or distracted by the object referred to. If Nim was to be taught the sign *cat*, it would be easier for him to practice making the sign *cat* (placing both index fingers below the nose and drawing them apart so as to rub the fingers over imaginary whiskers) without the disruptive presence of the animal itself. Later it would be relatively easy for him to form an association between the sign *cat* and an actual cat. This last stage of training was called "expression." During that stage Nim was required to make the desired sign in an appropriate context. He would be required to sign *drink* in order to sip some coffee, red when shown a red flower or a red ball, and so on.

Carol stipulated that each sign be taught in the fixed sequence: reception, production, and expression. She also urged that no teacher move to a new stage until Nim had mastered the preceding one. The purpose of this rule was to insure that Nim would not attempt to express a sign that he had not first understood through his teachers' usage of that sign. This was why training in reception was always to precede training in production.

I felt that Carol's curriculum had lots of merit as a means of getting Nim to pay attention to his teachers' signs and as a means of motivating him to sign. But I also felt that Carol's method had to be qualified in some respects and needed to be supplemented by other techniques. I wanted Nim to learn to sign in order to please his teachers and not just to obtain food and drink rewards. Accordingly, I instructed Carol to use food or drink rewards only when teaching a sign related to those rewards. At other times I wanted Nim to become receptive to praise and other kinds of social rewards.

I did not regard the method of reception, production, and expression as the only way for a chimpanzee to learn sign language, nor did I believe that the stages always had to proceed in the strict order stipulated by Carol. Other chimpanzees had been taught to sign by molding and without a strict rule that comprehension must precede production. Moreover, I knew of no evidence that children always learn to understand a particular word before they speak it. I made this clear to Carol and other members of the project at our weekly meetings, both before and after Nim began attending his nursery school. My hope was that, once Nim's teachers had experienced what it was like to work with him in the classroom, there would be a good basis for discussing which methods worked best.

Nim was usually brought to and from Columbia by Carol. Typically his day began with the ritual of removing his outer garments. Carol or the teacher in charge of the first session encouraged him to remove them by himself. At the same time, the teacher signed about what was happening: *coat, hat, pants, out.* Diaper changing was the next part of the daily routine. During this operation, the teacher signed *diaper, dirty,* and *powder.* Dressed in a fresh diaper, short pants, and a tee-shirt, Nim was carried into the classroom where he was fed a large breakfast. Like all of Nim's activities, breakfast was broken down into a series of small steps. First Nim was let down by the teacher and allowed to sit on the floor. Before putting him down his teacher signed *down.* The teacher then signed *chair* and helped Nim climb into his high chair. While holding his bib, the teacher signed *napkin.* Each food (*banana, apple, juice,* and so on) was identified in sign language, as were the utensils used (*spoon, knife, cup,* and *bowl*). While Nim was eating his breakfast, his teacher signed *eat* and *drink* at appropriate moments. After breakfast the teacher signed *clean* while wiping Nim's face and the high chair. Before picking Nim up, the teacher signed *hug.*

After breakfast, the teacher tried to interest him in an activity that had been prepared beforehand. Almost every day a bag of toys was brought out. With a flourish, the teacher removed each item one at a time. The contents of the bag, which varied from day to day, included the usual assortment of playthings that might interest a one-year-old child: balls, stuffed animals, mirrors, books, boxes nested one in another, blocks, balloons, drawing materials, matches, flashlights, and so on. The teacher tried to interest Nim in the contents of the bag by using the dramatic devices one would try with children: peeking into the bag, looking surprised, scared, horrified, puzzled, and so on. As always, each item was identified in sign language before it was given to Nim. After an hour or two in the classroom, Nim was allowed to run around in the internal corridor or in the small "gym" across from the classroom. In these areas he had a choice of activities: he could run around in the hallway, play in a sandbox in his gym, climb on a branch suspended above the sandbox, or hoist himself up on a chinning bar wedged into the entrance way of his gym. One activity Nim could never get enough of was being pushed around the internal corridor in a stroller. While he played, his teacher would sign to him about what he was doing: *play, jump, go,* or *climb.*

Within a month of Carol's arrival on the project, Nim had learned two new signs, *hug* and *clean.* During the next half year he learned another six signs. He first learned to sign *dog* at home, mainly in response to Trudge, the LaFarges' dog. But once he learned to sign *dog,* Nim would produce the sign easily in the classroom, in response to a picture of a dog. The other signs Nim learned during the fall and spring semesters of the 1974–75 academic year were *down, open, water, listen,* and *go.* Though he learned these signs in the classroom, he had no problem using them at home. In the classroom it was easier to focus his attention, to get

A break from instruction in sign language

him to notice what his teachers were signing and to practice his own signs. But the LaFarge home provided a greater variety of opportunities to sign than the classroom, and he often used signs at home in more interesting ways.

The presence of Stephanie and Jennie was another reason for the richness of his signing at home. At that time, he was closer to them than to any of his other teachers. That he signed more readily and more creatively with them confirmed my belief that a strong social bond provided the best motivation for learning language. Also, neither Stephanie nor Jennie was as interested in signing for its own sake as were Carol and many of Nim's other teachers. It was not that Nim's classroom teachers didn't like to engage him, as Stephanie and Jennie did, in many intimate ways when they played with him. The problem was that the motivation of Nim's classroom teachers was different. Unlike Stephanie and Jennie, Nim's teachers proved themselves, not by how well they played with Nim or by how well Nim liked them, but by how well they taught him to sign. Since their time with Nim was short, they devoted most of their classroom sessions to advancing Nim's knowledge of sign language.

Once I became aware of the different ways in which Nim reacted to Stephanie and Jennie on the one hand, and to his classroom teachers on the other, I tried to arrange a schedule that would allow the classroom teachers to spend more time with Nim at home. That effort was quite unsuccessful. Nim's teachers often found it more difficult to control Nim at home than in the classroom. Secure on his own territory, he had the confidence to draw upon a large bagful of tricks that made it difficult for his teachers to work with him. One basic problem was the sheer

Relaxing with Jennie at home

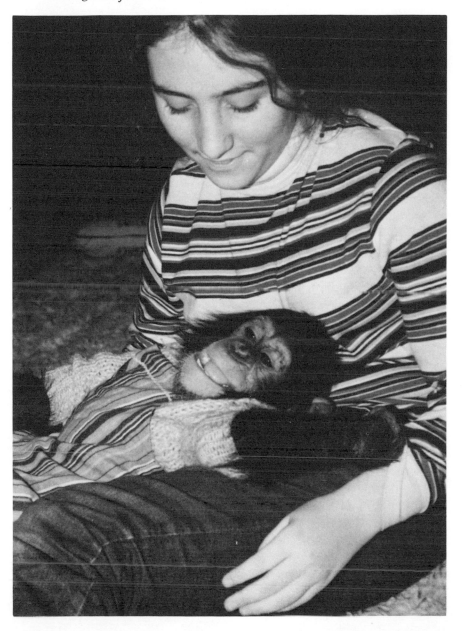

amount of space in which he could scoot around at home, and the many pieces of furniture he could jump on or hide behind. Just when a teacher thought that Nim had quieted down and was concentrating on what he was supposed to be learning, he would jump up on the piano, or race up the stairs and slide down the banister, or climb up on the bookshelves and jump down onto the waterbed beneath them. Matters became even worse if a resident member of the LaFarge household wandered by. Nim would often regard that person as a protector and escalate his mischief.

Nim's reaction to Stephanie and Jennie, and to a few other members of the LaFarge household who had established strong dominance over him, was quite different from his reaction to nonresident teachers. A mere look of displeasure from Stephanie sufficed to get Nim to stop what he was doing and to run over to her for a reassuring hug. Only by spending the many long hours that Stephanie and Jennie had spent with Nim could his teachers hope to obtain the kind of dominance over Nim that would enable them to work effectively at home. There were problems of space too. And WER LaFarge was becoming more and more annoyed about the minor damage Nim was inflicting on his home—chipped furniture, smudge marks on the walls, scratches on the floor—damage that seemed to occur most frequently when classroom teachers tried to work with Nim at home.

One of Nim's teachers was an exception to the rule that outsiders couldn't cope well with Nim in his home. Laura Petitto went out of her way to baby-sit for Nim when no one else was home at the LaFarge house. Laura also worked on her signing skills by taking advanced courses in sign language at the NYU Deafness Research Center. She managed all of this while attending college full time and earning her room and board by baby-sitting for the family she lived with. Perhaps her experience with the two young boys she looked after gave her the insights she had for dealing with Nim. Perhaps the work she had done at Ramapo as a research assistant on a project that studied the social behavior of baboons helped too. For whatever reason, Laura was able to develop an unusually close relationship with Nim, both at home and in the classroom. By the end of Nim's first semester at Columbia, Laura had begun to contribute strongly to his intellectual and emotional growth.

To outsiders, it appeared that Project Nim involved little more than looking after Nim and teaching him to sign. In a way this was true, but the extent of the work should not be minimized. Five days a week, for five to six hours a day, volunteer teachers worked with him in the classroom on what proved to be a rather grueling schedule. Yet the teachers complained less about their schedules than about other aspects of the project.

Looking at a picture book with Laura

At our weekly meetings, project members spoke of their experiences teaching and handling Nim and their problems with Carol's method. It wasn't long before our superficially harmonious and constructive discussions were disrupted by an undercurrent of tension. After a number of private conversations, first with Carol and with Stephanie and later with other project members, I became aware of problems for which there were no obvious solutions. It was important that Nim's teachers impose some degree of consistency in their social and linguistic demands, but I could not endorse a single formula or set of rules that was regarded *a priori* as the best method.

Carol argued that her method was not being applied broadly enough in the classroom or at Stephanie's house. Other teachers who disagreed with or who did not understand Carol's methods felt that they wanted more freedom to try their own techniques. Not having as extensive a liberal arts background as many of the other teachers, Carol was less prone to theorize or philosophize about what she was asked to do. She had lots of experience getting a job done in the most practical way possible, and she understood from me that I wanted a well-behaved chimpanzee who could learn to use sign language. That became her goal. In order to achieve it she applied what she knew from her practical experience with principles of behavior modification.

Other project members had been introduced to the principles of behavior modification in courses on behaviorism or learning theory rather
than through working in institutions. Accordingly, they were more attuned
to the theoretical context in which the principles emerged than to the
practical difficulties of applying them. Furthermore, they were aware of
the criticisms that had been leveled against using behavior modification
as an educational tool. Some had backgrounds in psycholinguistics and
cognitive psychology. Unlike Carol and her supporters, they were prone
to ask such questions as "What does a sign really mean to Nim?" and
"Did Nim really understand the sign signed to him by his teacher when
the teacher was using the method of reception?"

What began to emerge at our weekly meetings was a standoff between two factions. Kela and Bill, both of whom had volunteered to work
on the project partly because of their acquaintance with Carol, supported
her. So did Lisa Padden, a new volunteer from Hood College, in Maryland, who had worked with Carol at the Southbury School in Connecticut.
I and, to a lesser extent, Stephanie were the arbitrators between these
factions. Stephanie's impartiality began to be questioned when the debate
about how Nim should be worked with in the classroom extended to what
was happening to Nim at home. Carol felt that Nim wasn't disciplined
with enough consistency in Stephanie's house and that the demands for
signing there were too lax. She was not so much concerned about what
Stephanie and Jennie were doing with Nim as about other residents of
the household who were not so actively involved.

Once the question about the relationships between Nim's life at home
and his time in the classroom was broached, our sessions began to sound
like emotionally charged PTA meetings. But unlike most PTA meetings,
where the issue is usually why children are not taught more at school,
Project Nim meetings focused on whether Nim was being properly taught
at home.

How Nim should be disciplined for antisocial behavior such as biting
and screaming generated more controversy and emotion than any other
issue I can recall. Carol's position was that every time Nim bit someone
or screamed too loudly he should be placed in a small, dark enclosure for
a fixed period of time, a so-called time-out box. This is a standard procedure at various institutions, particularly those dealing with hyperactive
children. The point of this procedure is to communicate as consistently as
possible that whenever a child engages in antisocial behavior he or she
will be punished in a particular way. The form of punishment chosen was
one that briefly isolated the child from all social contact. Carol wanted
time-out boxes installed both in the classroom and at home.

From the outset, many members of the project, myself among them,
were opposed to using this form of punishment. It was not that we were

unconcerned about biting, screaming, or other antisocial acts. Bites from an infant chimpanzee (who has a full set of canines at ten months) are painful and become increasingly dangerous as the chimpanzee grows older. But it was not clear that the time-out box would be the best solution every time Nim bit someone. A practical objection to the time-out box was that it would be difficult, if not impossible, to use once Nim became strong enough to resist our efforts to put him in such an enclosure.

But my main objection to Carol's approach was that it was too simplistic. Nim would bite or scream for various reasons. To follow automatically each instance of antisocial behavior with the same punishment could make matters worse. He might bite because he was frustrated with his teacher: perhaps the teacher did not understand what Nim signed, or was too demanding of him, or punished him unjustly. He sometimes bit in order to establish dominance, particularly with new teachers or with some of the younger (and smaller) members of the project. And, at times when Nim was teething or ill, he may have bitten someone because of physical irritation.

I urged that before Nim was punished for undesirable behavior, a clear assessment be made as to why he engaged in that behavior, and his punishment chosen accordingly. There seemed to be available a wide range of forms of punishment other than the time-out box that might prove to be at least as effective and perhaps easier to administer. One alternative would be to ignore Nim for a fixed period of time. Nim could be left alone in the classroom without any toys or things to play with. As Nim grew older and learned to understand more language, I thought it might also be possible to sign to him about his antisocial behavior. The teacher could sign *You bad, Me angry you,* or *If you bite, I leave.* Nim had already learned that various types of praise following good behavior would lead to new activities and lots of attention from his teacher. I hoped that he would also learn that verbal admonitions would be followed by the loss of attention and of pleasurable activities that his teacher would normally have provided. If some form of corporal punishment were to be used, I felt that the kind of cattle prod that had been used at various primate centers would be an improvement over the time-out box. It would create less opportunity for struggle and would provide a more immediate punishment.

Rational discussions of these issues at our meetings were few and far between. When one teacher suggested a particular form of punishment, another would say, "You can't do that to *my* chimpanzee" or "Poor little Nim, how terribly these other teachers want to treat you." One of the teachers, who had apparently been institutionalized some years back, equated my suggestion of using a cattle prod with electric shock therapy.

After many discussions, I persuaded the group to adopt an eclectic policy. Intentional biting, scratching, hitting, breaking, and so on were considered to be unacceptable behavior that had to be punished accordingly. All teachers were asked to confront Nim and provide some kind of punishment for such actions. Those teachers who wanted to join Carol in using the time-out box were free to do so on an experimental basis. Other teachers were encouraged to discipline Nim in the manner with which they felt most comfortable, excluding the cattle prod or any other form of corporal punishment.

The issue of how Nim should be taught sign language was less emotionally charged than the issue of punishment. Still, discussion of this topic split the teachers into the familiar camps: for and against Carol's position. Aside from urging that her three-step method of reception, production, and expression be used more extensively and consistently, Carol argued that Nim should not be confused by inconsistent usage of signs by different teachers. For example, when he wanted a red cup, all teachers were urged to have him sign *cup* and not *red*. Predictably, the teachers who did not feel obliged to support Carol's position argued that strict rules regarding the usage of signs would not allow Nim to understand the meaning of those signs.

I felt that Carol's position that one and only one sign should be used in a particular situation was unreasonable. At best, it seemed to be a misguided application of the principles of behavior modification. The ideal that Carol was working toward was a chimpanzee who would mechanically produce a particular kind of behavior each time the teacher made a particular sign and also mechanically produce a particular sign when confronted with a particular situation. During receptive training, Nim was expected to jump down from the teacher's arms when the teacher signed *down*, take his coat off when the teacher signed *coat*, hang up his coat when the teacher signed *hook*, go to his chair when the teacher signed *chair*, sit in the chair when the teacher signed *sit*, and so on. Later, during expressive training, he was expected to sign *down* when he wanted to be released from the teacher's arms. Likewise, he was expected to sign *coat* and *hook* while removing his coat and hanging it up, *chair* upon seeing the chair, and *sit* upon sitting in the chair.

Children learn that it is appropriate to use a varied set of words in the same situation, such as red, cup, water, or drink upon seeing a red cup of water. Children also learn that it is appropriate to use certain words, such as big, give, or look, in different situations. Experimental psychologists have yet to resolve exactly how a child learns to make such abstractions, but, in any case, it is clear that a child can master a concept only if he experiences different exemplars of that concept. I felt that Nim could learn to make such abstractions too, but only if he was allowed to experience different word usage in different contexts. Restricting him to a single usage of every sign seemed to insure that he would not learn the concept associated with that sign.

There were other equally compelling objections to Carol's rule against confusing Nim through multiple usages of a sign. Too many exceptions could be found when trying to restrict a sign to one and only one context. Such a restriction might do in the case of concrete objects and people, but much of the vocabulary we tried to teach Nim included signs that, of necessity, referred to relationships, feelings, and actions. In each case, these signs applied to a widespread set of conditions. How could we possibly teach signs such as *open, in, more, hug,* and *sorry* in only one context? And even if we could, that would only tell us that Nim's teachers agreed on the contexts of the sign, not whether Nim perceived the same contexts.

As was the case in deciding how to punish Nim, I felt that an eclectic approach was needed to teach him how to sign. So long as different teachers followed general guidelines and did not produce training methods that negated one another, I saw no harm in using a diversity of methods in trying to advance Nim's intellectual and linguistic development. If one teacher wanted to use cups and bowls to teach Nim the difference between the names of those objects and another teacher wanted to use the same objects to teach Nim the names of their colors, I would approve of both approaches so long as they didn't result in what I judged to be a glaring inconsistency.

I urged all of Nim's teachers to use many of the techniques that Carol had introduced. Teaching Nim to be receptive to signs (or, as it is commonly called, to "comprehend" or to "understand" signs) was of obvious importance. And I believed that Carol's "production" technique was useful for teaching Nim the configurations of many signs. But I also felt it necessary to complement Carol's method with other techniques that admittedly compromised certain aspects of her method. For example, I felt receptive and expressive training need not occur in that order. And now that Nim was beginning to imitate signs, it should be possible to minimize or to skip over the production stage.

In an attempt to defuse the emotional reactions that interfered with rational discussions of socialization and sign instruction and to provide a factual background, I organized a weekly seminar to discuss assigned readings in three areas: language learning by chimpanzees, child development, and psycholinguistics. Unfortunately, whatever was learned at these seminars did little to ease the personal antagonisms that had developed. Carol felt that there was no proof in the readings I assigned that her method was wrong, and in a sense she was right. Proof that a particular method is "right" or "wrong" is rare in psychology. What Carol and those who supported her failed to see was that there was equally no proof that her method was in fact the best.

Had I started these seminars before opinion polarized, it might have
been possible to persuade all members of the project that we didn't know
enough to develop a rigorous method for teaching language to a chimpanzee. But after the *Sturm und Drang* of five months of debate, Carol
understandably required acceptance of her method as a vote of confidence. That I could not provide. In March 1975, Carol resigned from her
position as the administrator of Nim's instruction in sign language. She
wished to continue with the project as a teacher on equal footing with
the other teachers. This meant that I would have to assume the responsibilities she was giving up. I did not want to disrupt Nim's unstable
existence further by replacing Carol with a new teacher, even if we could
have found one on such short notice. And despite Carol's inability to
coordinate the efforts of all the teachers on the project, I felt that she was
a capable teacher who could follow suggestions from me and from other
teachers. I also felt that she could work more effectively when relieved of
the need to persuade the others of the merits of her method.

Aside from teaching Nim specific signs, Carol made many other last-
ing contributions to the classroom. She hunted around town for novel
educational toys, did research on the food we fed Nim, tried different
schedules for his meals, naps, and playtimes, and looked after a host of
other details that helped the classroom run smoothly. But though she was
effective as a teacher, it soon became clear that she had lost her heart for
the job. No doubt she was discouraged by being denied the opportunity
to apply her method fully and by the progress of other teachers who had
developed their own methods. For these and other reasons, Carol an-
nounced her resignation from the project, effective the end of May.

Despite the controversies that consumed most of our meetings, Nim
made significant progress in learning to sign. Between September 1974
and May 1975, a period during which Carol worked as a full-time teacher
and during which we had a more or less homogeneous group of classroom teachers, Nim learned eight signs (*hug, clean, dog, down, open,
water, listen,* and *go*). During the end of that period groundwork was
laid for many additional signs Nim was to acquire during the next few
months. Of greater significance was Nim's increasing tendency to produce
combinations of signs, for example, *banana there, me more drink, tickle
me Nim,* and *give me eat apple.*

Dwelling on Nim's linguistic progress was, unfortunately, a luxury I
could not afford. At the end of May 1975, I had to concern myself more
with the practical difficulties of continuing the project than with Nim's
signing. The end of the academic year brought problems that were becoming all too familiar. Student volunteers first had to reduce their work
on the project because of final exams. Shortly afterward they would leave
the project in order to take paying jobs for the summer. In a relatively

short period of time, new volunteers had to be recruited, interviewed, and trained. During May 1975, I was faced with additional problems. It was not clear who would replace Carol, where the salary for that position would come from, or where Nim would live. Not only was I unable to borrow any more departmental funds, I was also obliged to show by June 1975 how I planned to repay the funds I had borrowed for Carol's salary. A grant application to the W. T. Grant Foundation was pending. If that application were denied, I would have no way to pay the minimum expenses needed to continue the project.

Even a grant from the Grant Foundation, however, would not solve the problem of where Nim would live at the end of the summer. Both Stephanie and I agreed that Nim should move out of the LaFarge house as soon as possible. Nim had become too big to sleep in the hammock tucked away in the corner of Stephanie's dining room. He needed a room of his own, which was not available in Stephanie's house. And he was breaking too many things. No matter how well-behaved he was, the intensity of his play made matters constantly worse. His ability to jump on just about any piece of furniture or hang from any pipe or chandelier insured that a certain amount of breakage would continue. Nim also needed a more stable environment and a smaller group of caretakers.

It did not seem very likely that in one month I could find a replacement for Carol, money to pay that person's salary, a new home for Nim, and people to look after him there. Accordingly, I told Dr. Lemmon that there was a good chance I would have to end the project and return Nim to his institute. Dr. Lemmon was understanding and told me that I could always count on the institute as a permanent home for Nim.

In June I began to look for a new home for Nim. Since I didn't know if and when I would obtain support for the project, I could not consider privately owned property. Instead I asked Columbia University if they had any vacant houses. My preference was for a small dwelling in the neighborhood of the university that would provide ample space for Nim both inside and out, and that could also accommodate three or four students to look after Nim. The only property that met these requirements needed extensive renovation. Even if I had been able to pay for it, it seemed unlikely that the house would be vacated before January 1976. The next step was to consider property that was not within walking distance of the university.

The first place I was shown outside of the university neighborhood was the carriage house of a large mansion in Riverdale—a lovely area of the Bronx about twenty minutes by car from Columbia. Situated on the Hudson River directly across from the Palisades, it is one of the few remaining green areas of New York City. The carriage house and the mansion were part of an estate given to the university in 1964 by Edward C.

Delafield, a former president of the Bank of America and a serious amateur botanist. Mr. Delafield hoped the university would use his estate as a botanical field station. The Delafield Estate, as it came to be called, was admirably suited for this purpose. It consisted of thirteen beautifully landscaped acres. The grounds contained a rock garden with many rare species of flowers and shrubs, a rose garden considered the finest of its kind in the East, handsome specimens of many varieties of trees and plants, and dozens of varieties of rare and exotic plants in three separate greenhouses. All that this wonderful estate needed was a botanist who would turn it into an experimental field station. From the viewpoint of the Columbia biology department, however, the timing of Mr. Delafield's gift was all wrong. Molecular biology, not botany, was their main focus. Thus the estate never served a scientific purpose. For a brief time it had been the residence for Grayson Kirk, a former president of the university, but since 1973 it had stood vacant except for caretakers. Maintenance costs were rising, and the university, seeing little likelihood of the estate's being used for the purpose intended by Mr. Delafield, had put it up for sale. For the moment, however, the sale was tied up in legal technicalities, so Delafield was available for the use of Project Nim.

When I first saw Delafield, with its sweeping driveway and classical

Delafield

portico, it reminded me of Tara in *Gone With the Wind*. It had been built in 1895 in the Georgian style, on the crest of a hill with views of the Hudson to the west and north. I inspected the carriage house and concluded that with a little work it could be made livable for Nim and a small staff. Then, on a whim, I asked if I could look at the mansion itself.

It was as impressive inside as out. A modest-sized reception area, whose walls were covered with murals of old New York, belied the spaciousness of the rooms to the left and right. The northern wing of the first floor housed a library and a living room. The living room looked especially cavernous because it had no furniture, but it was really bigger than many two-bedroom apartments (approximately thirty-seven by twenty-three feet with an adjacent twenty-three-by-eighteen-foot sun porch). The southern wing of the first floor housed a formal dining room with an attached sun porch, a large kitchen and pantry, a breakfast room, and a laundry room. On the second floor were four spacious bedrooms. The master bedroom was actually a suite consisting of a sitting room, a terrace overlooking the river, a dressing room, a bathroom, and a twenty-five-by-eighteen-foot bedroom. The third floor contained an additional bedroom and two servant's rooms. The more I saw, the more I felt that this might be the perfect home for Nim. The twenty-one-room mansion had much more space than the project could possibly use, but the carriage house was claustrophobic.

Shortly before coming to see Delafield Estate, I had learned that the W. T. Grant Foundation would make an award of $10,000, commencing July 1, to cover our minimum expenses through the end of 1975. Since I had two larger grant proposals pending, one with the National Institute of Mental Health and the other with the National Science Foundation, I felt there was a good chance that the project could finally proceed on a solid financial basis and be in a position to pay for accommodations. I asked the university whether I could rent about half the area of the mansion, and to my great surprise and pleasure, they agreed. Their only stipulation was that I be prepared to vacate Delafield if they were able to sell it. But I was assured that if Delafield were sold, the university would try to find me similar facilities on property they owned in Westchester and Rockland counties. Since a sale did not seem imminent and since Delafield was such an ideal place to house Nim and his caretakers, I was happy to sign a lease that stated that I would vacate the premises on ninety days' notice.

The space was clearly more than I needed for Nim and four resident caretakers. In order to minimize my responsibility for the care of the mansion and to reduce fuel costs, I sealed off the library and the living room, half of the first floor. What remained still exceeded the project's needs, but after the cramped quarters of the LaFarge house, it was nice to have too much space.

After learning about the funding from the Grant Foundation, I asked Laura Petitto, who was outstanding among the volunteer teachers, whether she would assume Carol's job as Nim's full-time sign teacher. Laura had already made plans to go to graduate school in psychology, but she was also aware of her skill in working with Nim and believed strongly in the future of the project. Fortunately for Nim and for the project, Laura agreed to accept the only full-time job on the project and to postpone her graduate education until the following year. To show my appreciation of that decision, I offered her the master bedroom.

I was also able to persuade Amy Schachter and Walter Benesch, two volunteer teachers who had joined the project during the spring of 1975, to move to Delafield. Both Amy and Walter had proven themselves effective teachers in the classroom, and they also impressed me as conscientious individuals whom I could trust to look after Nim. The rooms I was able to offer to Amy and Walter were not quite as spacious as Laura's, but they were bound to be *some* improvement over the typical college student's room.

My announcement that Delafield would be Nim's new home raised the morale and the spirits of everyone on the project. In no time the new residents of Delafield had cleared away the many years of dust and grime that had accumulated in their rooms and furnished their living areas with simple but charming pieces of furniture that they borrowed or built themselves. Laura framed the enormous red marble fireplace at one end of her bedroom with large, colorful pillows and tall bamboo stalks that were growing wild in one corner of the estate. Throughout the rest of her room she placed many exotic plants that the caretaker provided from the Delafield greenhouses. The bamboo, the plants, and the closeness of the trees to the four large windows of Laura's bedroom helped create the impression of being outdoors.

Amy's room, on the top floor, had a wonderful view of the Palisades and the Hudson, both framed by an opening of the trees that led down to the river, and a beautiful fireplace decorated with Delft tile. The mantelpiece was laden with objects that Amy had collected while working at the Bronx Zoo: purple peacock feathers, ostrich eggs, and snake skin. Back on the second floor, Walter's long, narrow room exuded an air of mystery. Through the dim light one could see evidence of his interest in anthropology and the occult: shrunken skulls, African masks, ancient Mexican artifacts, and, above his fireplace, the red velvet robe and the black crossed swords signifying his membership in a "secret society," which he occasionally spoke about but never identified.

Much planning and work went into preparing Nim for the move to his new home, but there was no guarantee that the absence of Stephanie and her family and the abrupt change to a new location might not have a traumatic effect on Nim. In order to minimize the disruptiveness of this move, I asked Stephanie to allow Amy to live in her house during July and the first part of August, and I asked Walter to increase his classroom sessions with Nim from two to four a week. As Nim began to feel more comfortable with Amy, Stephanie began to spend less time at the house and more time on her farm in Rhode Island. By the time Nim was ready to move, he was being cared for exclusively by Amy, Laura, Walter, and a few other volunteers.

At Delafield, Laura supervised a group of carpenters who had to complete a long list of renovations in Nim's part of the mansion before he was moved. They built a swing in what used to be the dining room, covered the lower portions of the French doors leading from the dining room to the sun porch with screening material so that Nim couldn't break the glass panels, built a loft bed for Nim in a small room on the third floor that used to be a servant's room, encased the fragile pipes that ran along the kitchen ceiling (pipes from which Nim would surely swing if he could), installed hooks to secure kitchen cabinets, and so on. Most of our initial attempts to chimp-proof Delafield ultimately proved inadequate, but at least we had begun what turned out to be an ongoing process of protecting an estate that had never been intended to accommodate a growing chimpanzee.

Nim did not know it, but within the space of six weeks he had been saved from the fate of being returned to his birthplace in Oklahoma. Instead of being returned to the Institute for Primate Studies, Nim was to become nothing less than the lord of a manor.

6

Salad Days

How Nim himself experienced Project Nim must remain, at least for the moment, a matter on which he cannot comment. From my point of view, however, the project went through several distinct periods. The first began when Nim arrived in New York in December 1973 and lasted for about eighteen months. During this period Stephanie was the most important person in Nim's life. The second period began when Nim was moved from Stephanie's house to Delafield during the summer of 1975. The start of the second period also coincided with two other important developments: a shift in responsibility for Nim's care and sign instruction from Stephanie and Carol to Laura Petitto, and the award of the project's first grant.

The time that followed Nim's move to Delafield was not particularly tranquil. Continuing worries about money led to a number of changes and to uncertainties about the future of the project. Looking back, however, it is plain that these were the salad days of Project Nim. The new master of Delafield blossomed into a proficient signer. Laura developed into an extraordinary teacher, and a highly talented group of volunteers amazed me by the long hours they spent teaching Nim to sign and performing the extensive data analyses needed to show just what Nim was signing about. During this period there was an unusually good mesh of the personalities within the project. Good relations between Nim and his caretakers, both in the classroom and at Delafield, led to important advances in Nim's social and linguistic development. At the same time, everyone on the project pulled together toward the goal of having our efforts recognized by a major granting agency.

By the time Nim was eighteen months old, just prior to his move to Delafield, I had ample evidence that the goals of the project were realistic. Nim had become well socialized, and his use of sign language had developed in many directions. Gratifying as this was, I was still frustrated by the failure of two of my original expectations to materialize. The first was funding. It was clearly much harder to get support for this kind of research than I had anticipated. Equally vexing was Stephanie's unwillingness to assume responsibility for teaching Nim to sign.

The second of these problems was solved thanks to the efforts of a remarkable young woman who, at the age of nineteen, agreed to serve both as Nim's main mother and teacher. Of all of Nim's teachers, there is little question in my mind that Laura was the most effective and influential in establishing methods for teaching Nim to use sign language and for documenting his usage of that language. Indeed Laura's performance during her two-year tenure on Project Nim suggests to me that the success of this type of research may, more commonly than is recognized, depend on the efforts of a single individual whose effectiveness is hard to predict from prior experience.

At the time Laura joined the project she was a college junior with a respectable record of research experience on projects supervised by her professors. She knew no sign language and had never worked as a teacher of normal or abnormal children. Yet her performance surpassed that of college graduates and graduate students, many of whom knew sign language and who were in special programs for working with disturbed children. If Laura hadn't performed so well the duties she assumed in June 1975, I doubt that Nim's interest in sign language would have developed much beyond what it was at that time. Instead it matured in so many ways that, when Laura left fifteen months later, there was enough data for me to begin to ascertain whether Nim could create and comprehend primitive sentences in sign language.

Laura's leadership in reorganizing the classroom galvanized the other teachers into adopting a more positive attitude toward teaching Nim. For the first time they shared a common outlook on how to work with him. Whereas much of the teachers' energy had been previously invested in defending their positions against opposing points of view, there was now a free exchange of ideas and an atmosphere of genuine cooperation and sharing.

On August 10, 1975, following a typical day in the classroom at Columbia, Laura drove Nim to his new home. She could sense Nim's apprehension even before the house was in sight. No doubt Nim picked up some of the excitement Laura was transmitting, and he probably realized that he was not being driven back to Stephanie's house. (That he was able to recognize that route seemed abundantly clear when I drove Nim past Stephanie's house some nine months later. A few blocks away Nim began to hoot. When we turned into Stephanie's block, Nim's excitement intensified further. As we drove past Stephanie's house, Nim stood up in his seat and pointed to the house in which he had spent his first eighteen months.)

Laura parked her car just outside the fence surrounding the Delafield estate and pointed to the mansion in plain view at the head of the driveway. She also signed to Nim that this was his new house. Nim seemed to sense that this was no ordinary event. His hair went erect, and he tried to cling to Laura. Instead of driving Nim right up to the door of his imposing new home, Laura decided to walk with him up the long driveway, to see if his curiosity would draw him to the house. To a large extent Nim behaved as Laura expected, but during the walk to the house, his hair remained erect. With one hand he clung to Laura's leg as if for dear life; with the other hand he signed *hurry* and slowly guided Laura toward the house.

Inside the house, Nim had to be carried from room to room. This was in sharp contrast to his behavior in Stephanie's house, where one of the problems was that he ran around too much on his own. After walking with Nim through his new quarters, Laura was joined by Amy for the first of many dinners on the sun porch overlooking the rose garden. Whenever either Laura or Amy strayed too far, Nim cried and jumped into the arms of whoever was closest.

It took Nim almost a year before he was as tolerant of separation from one of his caretakers at Delafield as he was of separation from a member of the LaFarge household. Week by week he allowed Laura, Amy, and Walter to increase their distance from him without showing signs of distress. But it was not until the following summer that he tolerated being left alone when he was awake. Even at bedtime, he was reluctant to part with his caretaker unless he was very sleepy.

The space in the Delafield mansion that was used by Project Nim was divided into two overlapping areas: Nim's quarters and his caretakers' quarters. Each caretaker had a bedroom and a bathroom on either the second or the third floor of the mansion, and those areas were strictly off limits to Nim. In order to insure that Nim did not visit his caretakers in their rooms, special doors and locks were installed at the two points at which he might have entered their area.

Nim had five rooms for his exclusive use. On the third floor he had a bedroom, a small playroom, and a bathroom. These rooms had previously been servants' quarters. On the first floor the large formal dining room had been converted into a chimp gym. Another servant's room, up a small flight of stairs behind the breakfast room, was converted into a classroom for occasional use, though Nim went to his old classroom at Columbia nearly every day. The kitchen, pantry, laundry room, breakfast room, and the sun porch behind Nim's gym were "common" rooms, in the sense that they could be used by both Nim and his caretakers. In most instances, however, these rooms were common only to Nim and the caretaker he was with at any particular time. When more than one caretaker was present, Nim loved to "socialize" so much that he was often more difficult to control. For that reason only the caretaker responsible for Nim was allowed to use the common rooms while Nim was around. There were, of course, exceptions, such as Sunday brunch, when Nim ate with all of his caretakers.

Going to bed was the most difficult part of Nim's first day at Delafield. His bed was a loft about four feet off the ground lined with carpet and outfitted with a blanket, a pillow, and one of Nim's favorite dolls. This new bed was spacious when compared to the small hammock in which he used to sleep in Stephanie's house. However, Nim's first night at Delafield was spent on the floor of the bedroom cuddled between Laura and Amy. On subsequent nights either Laura or Amy stayed with him. By the third night Nim allowed himself to be put in the loft but only if Laura or Amy climbed in with him. It was not until a week later that Nim could be left alone in his bedroom.

Nim often cried even after he allowed his caretaker to leave his bedroom. At times his cries subsided shortly after his caretaker left. At other times he woke up hours later and cried as if having a nightmare. Even though it was painful to hear his screams, I thought it wise to ignore them lest Nim learn that screaming was a way of making his caretaker return. But sometimes Nim's screams were so loud and insistent that we feared that he might choke and pass out. On these occasions I suggested that someone look into Nim's room to see that he was breathing normally. I am happy to say that he always was.

Within a month of Nim's move to Delafield, his screaming, both when he was put to bed and during the night, subsided. However, at intervals, sometimes months apart, Nim would awaken the household in the middle of the night with ear-splitting screams. We could never be sure what it was that caused them—a frustrating day, noise from a thunderstorm, a bad dream. Whatever the reason, it was impossible not to think of Nim as a small child crying out for someone to comfort him.

Nim's adjustment to his new home was slow but steady. Though he gradually allowed his caretakers to increase their distance from him, he would not allow them to leave his sight. If they did, he would throw a temper tantrum, screaming loudly and flailing around on the ground. Like the screams from his bedroom, his screams during a temper tantrum were far louder and more piercing than anything one might hear from a child.

The fact that Nim got upset when he was too far away from his caretaker became the basis of an effective way of disciplining him. As he became more adjusted to his new home, he often mischievously opened cupboards and drawers and threw their contents on the floor. Sometimes he ran around the kitchen challenging his caretakers to catch him. In his classroom at Columbia he climbed up the door frame and then jumped down on his teachers. Hitting Nim was mildly effective in these situations, but that of course presupposed that the teacher could catch him. Even when it was possible to grab him and spank him, he tried to convert the spanking into roughhousing, an activity of which he never seemed to tire.

A more effective technique of disciplining him was to walk away, preferably while signing something like *you bad* or *I not love you.* Instantly he would stop what he was doing and run toward his teacher, often signing *sorry, hug,* or both. For many months this procedure sufficed. But as Nim got older and more independent, the *sorry-hug* routine became more of a game than an expression of true remorse. Too often his behavior was just as mischievous after he signed *sorry* and *hug* as before. His caretakers then began to delay their response to his request for reassurance. Nim's reaction to this delay was to throw a temper tantrum. Painful as it was to watch these tantrums, it was sometimes necessary to communicate to Nim that we really didn't enjoy his bad behavior. After a tantrum Nim's behavior almost always changed dramatically for the better. The exceptions were instances in which a teacher reassured Nim too quickly.

Nim's temper tantrums were not confined to situations in which he was being disciplined. At times they were Nim's way of expressing outrage that he could not have his way. Like a child, Nim seemed to have learned that a loud outburst of emotion exerts a strong influence on an adult's behavior. As exhausting as these tantrums must have been for him, they usually produced the desired results. Consider two rather simple examples.

A few days after Nim's second birthday, Laura toasted two bagels, one for herself and one for Nim. After finishing his bagel, Nim came over to Laura and signed for more of hers. When Laura refused to give him any, Nim went into a tantrum that she described as follows:

> His lips drew together into an open mouth. A low hoot followed, gradually rising to a loud screeching sound. His lips pulled back tightly exposing his teeth with a very frightened and alarmed look. Nim never tried to bite during a tantrum. Most of the harm done was to himself. He pulled at his hair, threw himself onto the floor, onto objects; sometimes he hit himself until he was in such a frenzy that he began to choke. Occasionally, it appeared that even the choking was deliberate.

Taken aback by that display of emotion, Laura reluctantly gave him the remaining bagel half.

A few days later a small group of people visited Nim while he and Walter were in the kitchen. Nim's guests were Ronnie Miller and Steve Hasday (both volunteer teachers), Laura, and Amy. As always, Nim enjoyed being the center of attention and having so many people to play with. Also, since Nim had moved to Delafield, he had shown an increasing desire to participate actively in social groups. At one point during the play, Walter finished making Nim's dinner and took him into the breakfast nook adjoining the kitchen to feed him. Ronnie, Steve, Laura, and Amy remained in the kitchen. About fifteen minutes later they were startled by a high-pitched scream. After the screaming had persisted for a few moments, Amy and Laura entered the breakfast nook and observed Nim standing in his chair while Walter, in front of him, was signing *sit*. Each time Walter signed *sit* Nim screamed and threw another tantrum. It was clear that Nim didn't want to be left out of whatever was going on in the next room. On this occasion the problem was solved by the departure of Nim's guests.

As Nim became more accustomed to Delafield, his interactions with his caretakers broadened in many ways, and that in turn enhanced his language development. Nim tried to include himself in many activities— cooking, cleaning dishes, washing clothes and cars, and fixing things— much more than he had at Stephanie's house. There were several reasons for Nim's developing interest in sharing chores with his caretakers; perhaps the most important were the absence of children with whom he could play and the fact that he learned the pronoun *you*.

Nim learned to sign *you* in January of 1976. Work on that sign had begun months earlier, shortly after Nim was moved to Delafield. His interest in sharing undoubtedly contributed to his understanding. Laura's diary tells how Nim learned to sign *you*.

> When Nim was very young, much of his attention was focused on himself and his immediate surrounding environment. Much of Nim's "world" moved from the outside to him. He grabbed things, ate things, and seemed to think that objects existed merely for his pleasure.
>
> It was relatively easy during this stage to mold the *me* sign. It is my feeling, however, that Nim did not come to understand *me* fully until the concept of *you* was successfully developed. This occurred over a period of 3 months during the Fall of 1975 (Oct.–mid-Dec.). At this time I engaged him in very focused *you-me* "concept play" (as I called it). In much of this activity the focus was on activities which allowed us to share—exchange things—*take turns*. Some activities included ball throwing, making tea, *you-me* ride in wagon, etc. It was not long after this that the frequency of *me* skyrocketed.
>
> Soon after *me* solidified . . . came the appearance of the *you* sign. It occurred during a snack break that Nim and I took after some roughhousing with the toy wagon. I began eating the yogurt, realizing that I was hungrier than I thought. I appar-

ently ate too much. Nim seemed outraged. He began slapping his chest in the *me* sign and then suddenly he reached over and began poking me precisely in the center of my chest. This was the beginning of the *you* sign. It was not long after this that he was able to shift from touching the other person's chest to merely pointing to him.

Nim appeared to have finally recognized the existence of an outside world, separate from himself, and his signing changed accordingly. Once he had mastered the meaning of *you*, he began to sign sequences such as *you tickle* (as a request to be tickled), *you me* (as a request for his care-taker to give him something), *you clean* (to ask someone to clean a dish or the counter top), and *you open* (to ask someone to open a door or a box).

One interesting use of *you* was reported by Laura and Patti Sparveri, a volunteer teacher. They had both taken Nim for a walk on the Columbia campus, and while they were sitting in a central area, a crowd of people gathered around them. The crowd included a group of three children, who kept their distance from Nim. Nim had been eyeing the children for some time, and it was clear that he wanted to play with them. He signed *give me* and *come* four times to the children. Wanting to avoid a situation in which Nim might hurt a child, Laura attempted to distract him by taking out his ball. This worked for a while. Nim signed *ball* a few times, and Laura gave him the ball. Unexpectedly Nim turned toward the chil-dren, signed *you* a few times, and then took the ball and threw it to one of the children. That child in turn threw the ball back to Nim. Nim again signed *you*, picked up the ball, and threw it back to the same child. This game went on for some time. Nim was trying to reach the child both through the use of sign language, by addressing the child as *you*, and through a medium of exchange. As far as I know, this was the first time that Nim engaged in an activity where the action proceeded from him outward to the external world. At this stage he apparently had enough trust that he was going to get back what he gave away.

Two incidents that Laura described in her notes should illustrate her special relationship with Nim and how he was learning social skills other than sign language. On one occasion, Laura left him on the diaper-changing table by himself for a moment while she went inside the class-room to prepare his next activity. With the stealth of a jewel thief, Nim slipped into the observation room. At the beginning of Laura's session, Nim had apparently heard Carol, his previous teacher, preparing a bowl of cereal in the observation room for his lunch. (It was a common practice for the teachers to help each other perform various tasks that were diffi-cult if not impossible to complete in Nim's presence.) After preparing

Nim's bowl of cereal and observing Laura through the one-way window, Carol had left the observation room. Nim could easily have heard Carol's departure. He had long since learned that the sounds of the doors to the outside corridor and to the observation room were reliable cues as to the comings and goings of his teachers.

Laura emerged from the classroom and went into the observation room to get the bowl of cereal she planned to give to Nim. When she couldn't find the bowl, she immediately suspected that Nim had done something with it, but she had no evidence. What puzzled her was that the bowl itself was missing. If Nim had simply eaten some of the cereal, the bowl should still be there. Meanwhile, Nim was just where Laura left him: lying on the diaper-changing table. It was Nim's exaggerated look of innocence that prompted Laura to grab him and to bring him into the office, sternly signing, *where bowl!* Nim gave Laura a puzzled look but nevertheless showed that he understood Laura's question by looking around the office as if to help her locate the bowl. Laura became more and more angry. Finally, she exploded, grabbed Nim by the hand and pulled him up on the changing table. With one hand raised as if to strike him, she again demanded to know where the bowl was. Nim must have sensed that he was in trouble. Whimpering, he took Laura's hand and led her to the sink immediately adjacent to the changing table. In a corner of this fairly deep sink, Laura saw the half-finished bowl of cereal.

At the end of another session, Laura was changing Nim's diapers before turning him over to Carol, his next teacher. Carol had knocked on the door to make sure that all was clear in the internal corridor. When Nim and his teacher occupied that area, the person at the door was told to wait until they could get back inside the classroom. The purpose of this rule was to guard against the distracting presence of a new teacher while the current teacher was changing Nim's diapers. Laura told Carol that she would be ready in a few minutes and proceeded to clean and change Nim. She then went back into the classroom to pick up the materials with which she had been working. Nim quietly got down from the changing table and walked toward the door to the outside corridor. Laura observed Nim with a watchful eye. She was sure that he would try to open the outside door and run into the corridor. She was not very worried, however, because even if Nim could open the spring-loaded hook that secured the outside door, it would take him a good bit of time to do so. To Laura's surprise, Nim did the opposite. He inserted the end of the obsolete chain lock into its receptacle as if to lock the door. Nim did not want Laura to leave! That this was not idle play was confirmed by Nim's subsequent behavior. When Laura finally opened the door, Nim physically resisted being transferred to Carol and whimpered for some time after Laura closeted herself in the observation room.

Much of Nim's waking life at Stephanie's house had been taken up with chase and tickle games initiated sometimes by Nim and sometimes by Jennie or Josh. But he was excluded from the work of the house, and because of the heavy demands of the human occupants on the kitchen, he was regarded as a nuisance when he played with various appliances and utensils. At Delafield, the kitchen was often Nim's own territory. Other residents were usually denied access when Nim and his caretaker were there, and Nim could focus on what his caretaker was doing with fewer distractions and more encouragement than at Stephanie's house.

Consider how Nim came to wash dishes. At first he would watch his caretaker perform the task while being carried in the caretaker's arms or sitting on the counter top next to the sink. Later he began to sign *give* as a request for the sponge. When given the sponge, he would make a few dabs at wiping a dish. Gradually Nim injected himself into the dishwashing routine until he did everything himself, chimp style. Laura described the process in her diary:

> With intense interest he would take the dish and put it into the sink, turn on the water, pick up the detergent, put it on the dish, and rub vigorously. This would continue for 20-25 minutes and he would take all of the dishes that we had just finished washing and put them all back in the sink, wash each one and pile them up again. During this time he exhibited fascinating facial expressions. They were very similar to an expression of alarm, in which his lips would be pulled back exposing his teeth. But he was clearly not alarmed, or excited or angry. He showed a very intense focused interest on what he was doing. His facial expression was accompanied by a very relaxed body, in that his hair wasn't up and he was sitting comfortably on the side of the sink.

Nim enjoyed sharing household chores and signing about them so much that he acted like a brat when his teachers wouldn't let him help. Excluded from activities in the kitchen, he would open cupboards and drawers and throw their contents on the floor, knock over dishes drying next to the sink, or turn over the garbage pail in the corner of the kitchen. In the laundry room, if he wasn't included in the washing and drying of clothes, he would spread soap powder around the floor, turn on faucets, and scatter clothes about. When he was allowed to help, he was not only much better behaved but he also was likely to engage his caretaker in conversation. Often he would sign *give me* to his caretaker in order to obtain some clothes to put into the washing machine. At the washing machine Nim would sign *open,* and then, having helped to load the

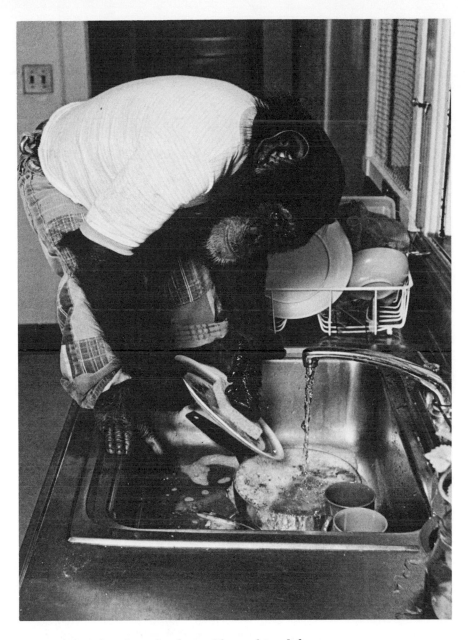

Nim loved to share the chores, like washing dishes.

machine, he would sign *wash* or *give wash* in order to get some soap powder to pour into the machine.

Nim especially liked to help prepare his dinner. Even when it took a long time, Nim showed remarkable patience and diligence in adding ingredients, stirring, and mixing things under the watchful eye of his

Another favorite activity: doing the laundry

caretaker. For example, when his meal was spaghetti and tomato sauce he would help fill the pot with water, put in the spaghetti, pour a can of tomato sauce into another pot, and stir the sauce while it was cooking. With one hand curled around the neck of his caretaker, Nim held on for support and watched what was cooking very intently. With the other hand he performed the various jobs his caretaker asked of him. When the spaghetti and the sauce were cooked Nim helped to pour the sauce over the spaghetti. Throughout the preparation of the meal Nim was allowed to taste what was cooking, but the tasting was not nearly so important to him as the opportunity to help out.

Once when Laura was in a hurry to cook and didn't want to be slowed down by Nim, she ignored him and signed that he should sit and watch. Nim protested by throwing a small tantrum and retreating into the corner of the kitchen farthest from Laura. There he stared at Laura while she prepared his meal. At the precise moment when Laura looked at Nim and obtained good eye contact with him, he deliberately threw over the garbage pail. As soon as he committed this act of defiance he stared at Laura as if to ask her what she was going to do about it. Laura, who had been a bit flustered before cooking the meal, exploded.

She signed *angry* and approached Nim with a menacing look. For a few moments Nim sat on the floor alternately hooting and rocking. Then he stood up and put every item of spilled garbage back in the pail. To complete his penance he took a sponge and wiped what was left of the garbage from the floor.

In one respect, Nim's adjustment to Delafield was quite egocentric. He began to regard two areas as his territory: the kitchen and the downstairs gym. When Nim moved into Delafield the dining room was converted into a spacious gym. First the handsome parquet floor was covered with some shabby second-hand carpet. This was done to protect the floor from Nim and Nim from the floor. Across the middle of the room we built a sturdy wooden frame that supported a few ropes and a tire. The frame and its fittings, all designed by Laura, provided Nim with a fine place for climbing and jumping. Aside from the frame, a large pillow, and some toys, Nim's gym was completely bare. All that remained of the former splendor of Delafield's elegant dining room was a marble fireplace, a mirror, and French doors that led onto a sun porch. Through the doors was a view of the rose garden, the Hudson, and the Palisades.

Nim also used the gym and the kitchen as areas in which he tested new people. Just about every new candidate for working with him was put through an initiation in these areas. If Nim saw a new person enter the gym while he was in the kitchen, he immediately ran to the gym and tried to drive the intruder out. In the kitchen Nim often attacked a new person and beat his hands on the floor or on cabinets. In a chimpanzee's natural environment that gesture is often interpreted as a means of asserting territory.

Nim's toilet training was initiated by Laura about a month after he moved to Delafield. Since she had never toilet trained either a child or a chimpanzee, Laura's plan was mostly intuitive. She began by placing a small white plastic potty in one corner of the Columbia classroom. Nim showed fear of the potty just as he showed fear of other changes in the classroom. At first he carefully kept his distance from the potty, staying in the opposite corner of the classroom. Later he approached it tentatively, holding on to Laura's leg with one hand and jabbing the potty with the other. Each time Nim struck out at the potty he looked as if he was making sure that he had an avenue of retreat or that he could hide behind Laura in case the potty counterattacked. Laura's reaction to Nim's antics was utter calm. She neither encouraged Nim to go near the potty nor discouraged his sporadic attacks. At that stage of Nim's toilet training Laura was concerned only that Nim would learn not to fear the potty.

Within a week Nim began to ignore it. By that time Laura had asked Nim's other classroom teachers to leave the potty in the classroom during their sessions, too. At home, Nim's caretakers left it in conspicuous view in

Nim's downstairs gym and in his bedroom. It wasn't long before the potty became Nim's constant companion. Laura and the other teachers carried it around everywhere so that Nim would regard it as a natural part of his environment. Laura went so far as to take it in her car when she drove Nim to and from his Columbia classroom.

A few weeks after the potty was introduced, Laura removed Nim's pants during her classroom sessions. This seemed to make Nim anxious. He covered his genitals with his hands and seemed concerned about the sudden change in the rules. For more than a year his teachers and care-takers had reminded Nim to keep his clothes on. Now Laura insisted that he keep his pants off. Nim seemed reluctant to defecate or urinate on the floor. When he did it was after much longer intervals than would have been the case had he worn diapers. On one occasion Laura noted that he "held it in" for eight hours. Most of the time Laura could tell when Nim had to go to the bathroom. He would give Laura a troubled look, lower his eyebrows, clamp down on his teeth, and hold his lips tightly together. Then he would get down on all fours and remain motionless with his muscles tensed. Usually there was ample time for Laura to get Nim over to the hallway sink. If there wasn't and Nim went on the floor, Laura drew Nim's attention to what he had done and then pointed to the potty before cleaning up.

On one occasion, just before Nim was about to urinate on the floor, Laura led him over to the potty and pointed at it. To Laura's delight Nim stared intently at it, then jumped onto the potty and began to urinate in it. From that point on toilet training proceeded quite smoothly. Both at home and in the classroom Nim began to use the potty regularly. That behavior was reinforced by Nim's being invited to watch his caretakers use the toilet. Like a child, Nim seemed fascinated by the toilet's con-tents, and he delighted in being allowed to flush the toilet after his care-takers used it.

Within three months of the start of toilet training Nim used a potty or a toilet regularly. At night he used a potty that was left at the foot of his loft bed. As a result he no longer had to wear diapers. His only toilet accidents seemed to be caused by emotional excitement: fear that he was going to be punished for some transgression, displeasure upon being transferred to a teacher he didn't like, and so on.

An important step in Nim's adjustment to his new home was his will-ingness to work with various caretakers in a small classroom that was modeled after the one at Columbia. Like the Columbia classroom, the one at Delafield was a small room, about eight feet square. Laura had stocked it with toys, books, and other educational objects in which Nim had shown interest at Columbia, but otherwise it was completely bare. This arrange-ment enabled Nim's teacher to follow the practice established at Columbia of focusing on one activity at a time. With increasing regularity Nim would lead his caretaker to the classroom, particularly after a long workout in the downstairs gym.

In both the Columbia and the Delafield classrooms Laura established a way of working with Nim that contrasted sharply with Carol Stewart's approach. No longer was Nim treated as an institutionalized child who had to demonstrate various behaviors or else. It was not that Laura made no demands on Nim. Indeed her demands were as strong as or stronger than those of any of the other teachers, including Carol. But Laura's demands were tempered with a combination of praise and play that seemed much more human than the strict approach advocated by Carol. What Laura sacrificed in the kinds of consistency that concerned Carol, she more than made up for in capturing Nim's interest and respect, two attitudes essential for successful teaching.

Laura's approach to working with Nim was deceptively simple. Many observers who watched them through the one-way window commented that Laura worked with Nim in such an effortless and spontaneous way that the whole process seemed terribly easy. What the untutored eye could not see was how Laura anticipated what Nim's next response might be. Having observed Nim carefully for more than a year, Laura could sense when Nim was losing interest in an activity, when he was genuinely sleepy, when he feigned sleepiness because he was bored, when he was malingering, when he didn't work on a task because it was too hard, and many other nuances of mood.

Establishing a good relationship with Nim meant reading his moods accurately and reacting appropriately. Thus when Nim was getting bored looking at pictures, the teacher had to know what activity to switch to and how to make that transition without upsetting Nim. Instead of mechanically taking the picture book away and substituting a new activity, say drawing, Laura would question Nim as to what he wanted to do next. More often than not he would reply that he wanted to play in the gym next to the classroom or that he wanted to be tickled. Rarely would Nim sign that he wanted another book or that he wanted to draw. Laura didn't always gratify his wish immediately, but at least showed interest in his preference. If play or tickle wasn't imminent, Laura would sign *later*, a concept that Nim readily understood.

Before a session with Nim, Laura prepared a lesson plan for a wide range of activities and on signs associated with them. She rarely used all of the activities she had prepared, but it was important to have the materials ready and to have thought through what Nim was to do and what signs he was to make in each case. Laura's preparation enabled her to create an impression of spontaneity. Having noticed that Nim was bored with one activity, Laura might "suddenly" notice a box she had left on

the shelf. Her previous training as an actress helped convey that she had
indeed just discovered the box. Her expression, which Nim studied very
carefully, was an invitation for him to explore the box with her. Before
presenting the box to Nim, Laura might sign *Open?*, *What in box?*, *You
want box?*, and so on. Upon opening the box, she would look amazed at
discovering its contents—whether it was a doll, a book, sunglasses, a ball.
She would go through this type of routine many times. Each time she displayed fresh wonder even though she and Nim had "discovered" the
same object dozens of times previously.

I think that it was Laura's spontaneity, her ability to anticipate Nim's
reactions, and Nim's respect for her that made her such an effective
teacher. The good teaching chemistry between Laura and Nim provided
a helpful example to other teachers. Through her work in the classroom,
Laura was able to inspire many teachers who had reached dead ends in
working with Nim. She also spent many hours observing new teachers,
often spotting idiosyncratic problems that existed between Nim and a
particular teacher. For the first time the project had a truly effective
trainer of other teachers. Laura's success in helping other teachers was
enhanced by her selfless approach to her work. She was very patient with
other teachers' problems, open to their suggestions no matter how bizarre,
and not threatened by the possibility that a new teacher might teach Nim
something that she hadn't worked on herself. Indeed, she would often
describe to me, with delight, a new technique developed by another
teacher.

A vital aspect of Laura's approach to Nim was her objectivity. Unlike
many other teachers, who saw him as a cute infant or a lovable animal-
playmate, Laura took him for what he was: an animal who was the subject
of a research program concerned with the linguistic ability of a chimpanzee. The best way to describe Laura's attitude toward Nim is to quote
again from her diary:

> Nim and I had a complicated relationship. . . . This is very hard
> for me to put into words . . . I had a feeling about Nim which
> was very platonic. This is my work, I have a certain goal, and
> this is what I am going to do. Initially, I quite honestly did not
> love Nim. It wasn't until a year after the project began that I
> began to feel that I liked him. I never thought of myself as his
> mother, which is one reason why I resisted people saying, "Oh,
> look at your cute baby!" Another reason is that I firmly believe
> that Nim did not see me as his mother. I think that Nim had the
> most information about me, the most consistent information
> about me, and therefore he was able to know where he stood
> with me at all times. He also respected me in that he respected

certain boundaries that I consistently held. I think that the key to my relationship with Nim was consistency. He had a sense of who I was and the way I wanted the world to treat me. . . . I'm sure I transmitted this to Nim. I never thought he was "cute," *per se*. I wouldn't let my feelings for Nim interfere with our goals. I used them to enhance our working relationship . . . in liking him, I began to respect him. I was in awe of the similarities he had to humans, he was an endless reservoir of information—I wanted to help him reach his linguistic potentials so as to provide him with a channel to teach me more about himself. That possibility motivated me to work very hard to fulfill the goals of this project.

A very important thing was that I watched Nim. Although I hated to sit in that observation room training new teachers and watching him for eight hours a day, I gained an enormous amount of information by observing Nim. Ultimately I began to feel arrogant about reading Nim. I could look at him and tell you what he was thinking, feeling, what he did two minutes before, and what he was about to do two minutes into the future. I can also say that I understand him. There is now no behavior that Nim emits for which I could not write a flow chart stating his purposes, his feelings, almost writing what the cause and effect of the situation on him was. I don't want to say that Nim is not a spontaneous creature, but like all of us, he was a creature of habit and to a certain extent, much of what he did, although it was incomprehensible to new people, was really predictable.

Under Laura's stimulating and conscientious supervision Nim's use of sign language developed in many directions. The rate at which he learned new signs—almost two a week—was never higher. Nor was that rate equaled after the departure of Laura and the other original residents of Delafield. Of greater importance, Nim was combining signs with increasing regularity, and the length of his utterances was growing steadily. During the summer of 1975, Nim was signing many two-sign combinations such as *more eat, tickle me, brush baby*, and *give apple*. During the summer of 1976, Nim was signing combinations of three or more signs, for example, *me more eat, you tickle me, me brush baby*.

Just as a child's use of speech is but one expression of its ability to interact with its parents and friends, Nim's use of sign language was but one expression of his personal growth. But compared to a child's, Nim's development was influenced by a relatively large number of adults. In order to place in perspective just what he learned about sign language from Laura and his other teachers, it would be helpful to take a closer look at his personality, at what kinds of teachers worked with him, and at how they reacted to him. As we shall see in the next two chapters, the many fascinating interactions between Nim and his teachers provided the crucible in which his knowledge of sign language was forged.

7

Nim's Personality

Of all nonhuman species, the chimpanzee is widely regarded as the most likely to master rudiments of a human language. The anatomical similarity between humans and chimps is one reason why. Among primates, the chimpanzee's brain is closest in complexity and in relative weight to man's. The chimpanzee's size and weight are also closest to that of man. Similarities between humans and chimpanzees are not, however, merely anatomical. From a genetic point of view, chimpanzees resemble humans more than any other species.

Similarities of emotional expression and personality are as compelling and important as anatomical and genetic similarities. Just how genetics and anatomy constrain personality is an interesting question but irrelevant to Project Nim. In communicating with another species, one of the most immediate facts to consider is the similarity of its personality to ours. Were it not for the natural empathy between human and chimpanzee, it would be difficult, if not impossible, to communicate through language.

Anyone who has observed a chimpanzee in a zoo, in its natural environment, or even in a circus should be able to recall how easy it is to recognize human personality traits in the chimpanzee. As Darwin noted in his classic book, *Emotional Expression in Man and Animals*, a chimpanzee's expressions of affection, joy, anger, curiosity, playfulness, and other emotions have a distinctly humanlike quality. Few people have an opportunity to express their personalities to a chimpanzee. If they did they would discover, as did I and many members of Project Nim, that a chimpanzee is also capable of discerning the moods of human beings. Indeed, given the mutual sensitivities of humans and chimpanzees and the many similar ways in which they express themselves, it often seems surprising that special training is needed to teach a chimpanzee to communicate via a natural language.

Even as a newborn infant, Nim had a distinctive personality. Most salient for me were his impulsiveness and directness, both qualities similar to but stronger than their counterparts in human infants. Especially when Nim was small, it was hard not to think of him as a human infant. One could, of course, argue that it is easy to react to small infants of almost

all species as one does to a human infant. My experience with Nim, however, left me with the impression that Nim came closer to having a human
personality than any other nonhuman creature I have encountered. What
other creature would explore my face and clothing and play with things
in such a human manner?

As he grew older, Nim did, in fact, learn a certain amount of
duplicity. On a number of occasions he lulled me into thinking that he
was content to play with his toys, and I lowered my guard. Once I did,
Nim would grab a grocery bag, my briefcase, camera case, or whatever
else I might have been carrying and run off to play with it. Most of the
time, however, Nim expressed his feelings and intentions directly. Often
they seemed to provide a mirror of my own feelings, that not only revealed them in their simplicity but also amplified them. This was true
whether Nim was expressing affection, curiosity, fear, aggression, wonder,
or determination. When I think of Nim what often comes to mind is the
image of a humanlike creature whose eyes often express an extraordinary
intensity and wonder or whose mouth and lips are pulled back in a
strong display of delight.

Affection was the most irresistible of Nim's emotions. It should come
as no surprise that while Nim was affectionate toward members of a
rather large group of volunteer teachers, his strongest expressions of affection were reserved for the people with whom he felt closest. What was
surprising, even to the people who knew him best, was the sheer intensity
with which he expressed affection. If Nim was playing with one of his
part-time teachers and I came into view, Nim would instantly move away
from his teacher and run toward me. Often Nim greeted me spontaneously with a kiss, after which he groomed me. Nim also reacted this same
way to other members of his immediate group.

Nim was especially affectionate when greeting a favorite teacher
whom he hadn't seen for a while. I would often be the target of strong
outbursts of affection even after only a few days' absence. Nim would
greet me with a screech of delight, sign *hug* repeatedly, and make an
"oooh, oooh" sound before jumping into my arms. At times he seemed so
excited that I worried about the way he sank his teeth into the base of my
neck. Even though I was sure that Nim was not trying to hurt me, I wondered if he knew the effect that his "love nibbles" might have on my
human skin.

During the last year of the project I witnessed several moving
reunions between Nim and former caretakers that revealed aspects of
his personality I had not previously encountered. Nim's reunion with
Laura took place in his Columbia classroom during December 1976, three
months after she left the project to start graduate school. Because she
didn't know quite what to expect and because she was feeling ill, Laura

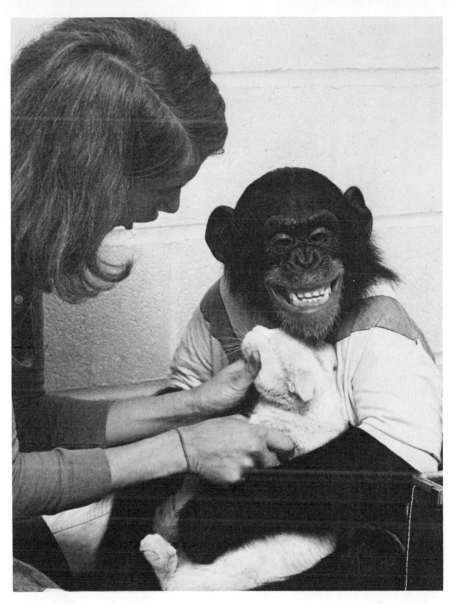

Nim's face was wonderfully expressive.

asked me to refrain from taking pictures, but she did allow me to watch her session through the one-way window of the observation room.

The only news that Nim had of Laura's return was transmitted by Dick Sanders, Nim's teacher during the session that preceded Laura's. I had asked Dick to show Nim a picture of Laura and to sign that she would be coming later. Nim kissed the picture but otherwise showed no special reaction. At the end of the morning session, Dick signed *Laura*

His eyes could show unforgettable intensity.

outside door! and *Laura come in here!* Nim studied Dick's signing very intently. This time he seemed more apprehensive, perhaps because of the urgency with which Dick was signing. When Laura opened the classroom door, Nim let out a screech of delight and jumped into her arms. His screams continued for some time. Inside the classroom Nim continued to jump up and down around Laura in a sustained expression of pleasure.

Once his excitement wore off, other aspects of Laura's special relationship with Nim began to surface.

Because she was ill, Laura tried to conserve her energy as much as possible. She began her session by sitting with Nim on the floor and grooming him, but Nim was not very cooperative. He was still too excited to lie quietly and be groomed. Instead he sat up, ran around a few times, and repeatedly signed *play*. Laura responded with as much energy as she could muster, but during most of the session she sat quietly and simply watched him run around the classroom. When it became apparent to Nim that Laura was not going to respond to his invitations to play, he began to test her. Testing was usually reserved for new teachers or for teachers who hadn't learned how to dominate him; if he tested an experienced teacher such as Laura, she would normally only have to look at him to remind him who was boss. Perhaps because he had not been with Laura for a few months, Nim seemed more daring on this occasion. Quite deliberately he pulled off his pants and threw them in the corner. While doing so, he watched Laura intently.

Instead of reacting, Laura sat calmly and watched Nim perform his antics. Nim was clearly puzzled. In the past, pulling off his pants would have resulted in a severe scolding, at the very least. As more time elapsed without a response from Laura, Nim looked more and more frightened. It was unusual that, by this time, a dominant teacher hadn't reacted to his transgression. Abruptly, Nim grabbed his pants and began to wave them around his head, fairly close to where Laura was sitting. Laura remained impassive in the face of this new provocation. She sat in the same position, feeling more amused than anything else by what was happening. Nim whirled around a few more times, holding his pants at arm's length so that they almost brushed against Laura's face. Again nothing happened. After a few more spins, Nim abandoned his efforts to provoke Laura and sat down quietly next to her. Again he stared at her intently. To me, it seemed as if he was trying to discern what Laura was thinking. Finally, she responded. She pointed to Nim's pants, which he was holding on his lap, and signed *In pants*. Nim obeyed immediately and signed *hug*. As soon as she smiled he went into her arms, clearly looking for assurance that everything was okay.

Even after a three-month absence from Nim's life, Laura retained a strong degree of dominance over him. If an inexperienced teacher had not responded to Nim's first test (in this instance, pulling off his pants), there would invariably have been a second test, which would have been more aggressive or defiant. He might have attacked the teacher, broken a toy, or defecated on the floor. Nim's reaction to Laura was quite different. Once he saw that he could not provoke her, he responded to her request to put his pants back on.

Nim saw Laura again at Delafield the following summer. This time Laura was feeling fine. She hid behind the trunk of a huge walnut tree on the back lawn of the estate while I walked with Nim along the back of the mansion, a route I followed frequently on our walks. Nim's attention was absorbed by the cameras hanging from my neck and by Susan Kuklin, a friend and a professional photographer, who had come along to take pictures, but he noticed Laura just as we had begun to walk past the walnut tree. A split second later he was in her arms screeching with delight. He and Laura spent a happy hour romping around the tree and signing with each other. Susan and I could barely keep up with Nim, but Susan did succeed in taking a few pictures.

Other teachers also experienced joyous reunions with Nim. On separate occasions, Marika and Walter came to visit Nim at Delafield. Each time Nim's response was one of total affection and delight, followed by playfulness and mild testing. From Nim's point of view his testing was probably just another form of play. As with Laura, it was fairly well contained. Once a teacher obtained dominance over Nim, only a slight reminder was needed to reestablish that dominance.

The most memorable reunion I observed occurred between Nim and Stephanie and her family. It took place at Delafield two days before Nim was returned to Oklahoma. Knowing how excited Nim gets when he sees a former caretaker, even from a distance, I took pains to insure that Nim had as little an inkling as possible of the presence of his visitors. I asked Stephanie and her family to sit quietly, around the corner from the main entrance to the house, so that Nim, who was playing in front of the house, would not see them. I then asked Dick Sanders, who was with Nim at the time, to walk with him toward the site I had chosen for the rendezvous between Nim and his first family. Nim was mildly excited when he saw me and signed *play*. That reaction was quite familiar. At that time I was visiting the house on a daily basis. Just about every time Nim saw me, he reacted the same way: a momentary round of playfulness, a quick hug, and no great disappointment when I put him down.

Stephanie had visited Nim at Delafield on two prior occasions. I had witnessed the most recent of these reunions. From what I saw and from what Stephanie told me about her earlier reunion, it was apparent that Nim was especially delighted to see her. But none of the reunions I had witnessed or heard about between Nim and Laura, Stephanie, Marika, or Walter had prepared me for the intensity of Nim's reaction when he suddenly came upon Stephanie and her family sitting in a semicircle near one of his favorite trees.

His response to Laura was as loving as ever.

Nim could not contain his joy at seeing his original family once more.

Nim smiled a smile the size of which I had never seen and shrieked in a way I'd never heard. At first he seemed too excited even to hug Stephanie. His smiling and shrieking continued for what seemed to be at least three minutes. During that time he sat down across from Stephanie. While looking back and forth at Stephanie, WER, Joshua, and Jennie, he continued to shriek, smile, and pound the ground with joy. Only after he stopped smiling and shrieking did he go to Stephanie and hug her. That hug was also interrupted by additional shrieks. Quite a lot of noise from a normally silent chimpanzee!

After spending about fifteen minutes with Stephanie, Nim went over

From left: Stephanie, Josh, Jennie, WER

to WER, Josh, and Jennie, and hugged each of them in turn. He then returned to WER and began to groom him. A few minutes later he moved over to Jennie, groomed her for a while, and then did the same with Josh. Nim's total involvement in hugging and grooming and in playing with the LaFarges was all the more impressive in that he seemed oblivious to the presence of a group of familiar and well-liked teachers who were watching from afar (Dick Sanders, Bill Tynan, Joyce Butler, Mary Wambach, and me). To Nim, it seemed as if nothing mattered but being reunited with his original family.

When he finished grooming Josh, Nim turned to Stephanie and

Re-establishing intimacy by grooming WER

her family and repeatedly signed *play*. In turn, each member of the family responded. Josh and Nim climbed a tree together. Later, Jennie and Nim chased each other around the base of the tree. Stephanie and WER also got into the act by running with Nim. Nim placed himself between Stephanie and WER, grabbed their hands, and pulled them to and fro around the grounds of Delafield. All of this seemed to be an incredible treat for Nim. Even after spending more than an hour with Stephanie and her family, Nim was still smiling. I had seldom seen even brief smiles of the kind that almost seemed pasted on Nim's face while he was with the LaFarge family. The only other times I

With Stephanie and WER

saw Nim smile that way were during the first few seconds of playing with one of his favorite cats and during the first few minutes of a re-union with a returning caretaker. I do not recall any situation in which Nim's smile persisted for so long or any time when he displayed such unrestrained joy.

I felt sad that the occasion of Nim's joy was a farewell, of which he had no inkling. Even though I didn't talk with Stephanie and her family during this entire episode, I could tell that they were also quite moved by Nim's reaction to their visit. When it was time to leave, neither Stephanie nor Jennie could hold back her tears as they walked away.

A few days later, in Oklahoma, I would also experience what it was like to say goodby to Nim.

Nim's strong expressions of uninhibited affection have often been misinterpreted by new teachers as evidence that Nim was always a sweet and docile young chimpanzee. It is true that Nim's first reaction to a new person was one of affection and curiosity. But that mood could, and usually did, change to a game of testing in which Nim reached into his bag of tricks to find a way of provoking his new teacher. That bag of tricks got bigger and bigger as Nim got older. Most of his tricks were variations on the theme "What can I do that I know is forbidden?" Nim's talent for creating mischief caused many a new teacher to wonder how a creature who seemed so gentle and affectionate could change so quickly into an unmanageable brat.

The number of ways Nim tested a new teacher was a measure of his socialization. Here is a by no means exhaustive list of things Nim would do to provoke his teacher: tear off his clothes; run away; turn over the garbage can; spill food on the floor of the kitchen; throw utensils and pots on the kitchen floor; destroy books, dolls, or toys; break a window; jump around in a car instead of sitting quietly in the front passenger's seat; have a toilet accident; grab a pocketbook or another of his caretakers' possessions.

Once Nim began to test a teacher who was not dominant, it was only a matter of time before he would escalate to more aggressive acts. Encouraged by his success in violating some rule, Nim would often attack his teacher. Undoubtedly Nim saw a teacher who could not control him as an opportunity to assert his own meager dominance. In the course of a few seconds, I have observed Nim change from a passive cuddly creature to a menacing little monster. When this kind of behavior occurred in the classroom, the only remedy was to replace the new teacher with an experienced one who would immediately discipline Nim for his bad behavior. When Nim acted up at Delafield, aggressive behavior could be diverted if the weather was good enough to be outside. The best strategy for a new teacher to follow when Nim proved too difficult to handle was to start a run-chase game or to simply allow Nim to play by himself in a tree. Of course, if the new teacher was going to sign with Nim, he or she would eventually have to learn how to engage Nim in activities that required him to sit still and attend to his teacher's signing.

Some new teachers were unable to control Nim even when they were willing to give him free rein to play. In those situations Nim seemed to sense fear or dislike on the part of his teacher. After a few sessions in which Nim and his teacher kept their distance from one another (literally, by the length of the lead with which Nim was attached to his teacher), Nim would work his way close to his teacher and attack. It proved impossible to prepare new teachers fully for the quickness and the ferocity of his outbursts of aggression. Despite extensive briefing about what to do when Nim attacked and extensive opportunities to observe others struggling to control him, new teachers often felt quite helpless when their turn came. Nim was amazingly quick and agile. Having defended themselves successfully against one attack, his teachers were simply unprepared for the speed with which he would attack again, usually in a different manner. Unfortunately for them, there was often too little time to defend themselves, and they would end up with some combination of a scratch, a bite, and torn clothes.

Disciplining Nim was an art that required more in the way of insight into Nim's moods and intentions and a good sense of timing than it did in the way of sheer brawn. Physical size and strength certainly provided an initial advantage. Nim's first reaction to a large person was to show respect and perhaps a little fear, but the advantages of size and strength were short-lived if the new teacher was unable to dominate Nim without recourse to physical force. Indeed, as Nim got older, physical force *per se* proved less and less effective. Two teachers who had exceptional dominance over Nim, Laura Petitto and Joyce Butler, weighed 119 and 115 pounds respectively.

Nim's aggressive behavior increased in frequency as he got older. In one sense that is not surprising. As he became stronger, he undoubtedly became more confident of his ability to hurt people, but I believe Nim's aggression increased mainly because of the necessity of introducing more and more teachers into his life. The sheer frustration of losing a caretaker with whom he felt close and having that caretaker replaced by a new person was sufficient to evoke Nim's anger and aggression. If it had been possible for him to have grown up with a small and stable group of caretakers, he would have experienced far fewer separations from his trusted caretakers and had far fewer opportunities to test his dominance through aggression.

Even though I exerted considerable dominance over Nim and was wise to most of his tricks, there were times he found new ways to test me. One memorable incident occurred during February 1976, shortly after I had begun a session with him in his Columbia classroom. That session was my first opportunity to spend more than a few minutes with him in more than a week. I could not be sure whether it was my recent absenteeism from the classroom or a bad session with his previous teacher or both that caused the cool reception I received from Nim. Even though he was well behaved, he was unresponsive to my attempts to engage him in various activities.

One of the first things I did was to see whether Nim wanted to use the potty. Since he could not yet be relied on to sign *dirty* every time he needed to use the potty, his teachers made a practice of pointing to the potty and asking *dirty?* In response to my question Nim walked over to the potty, pulled down his pants, and sat on the potty. I crouched down next to him and patted his back in praise of his good toilet habits. Having been away from the classroom for a week, I paid less attention than I should have.

In a flash Nim ran out of the classroom and opened the door leading to the outside corridor. When I had entered the classroom complex, I had carelessly left that door unlocked. This was not the first time that Nim had escaped from the classroom. Because of his curiosity about the hallways and offices of Schermerhorn Hall, I instituted a security system that called for all teachers to lock themselves into the classroom complex. Not only should I have followed my own directive to lock the doors to the classroom and the hallway, I should have also heeded the advice I had given to innumerable new teachers: never trust Nim. Particularly when a teacher was content with his behavior, Nim would sense that the teacher had lowered his or her guard. Unless he was reminded that the teacher was watching him very carefully he would usually find a way to exploit any lapse of vigilance.

Catching Nim was no easy task, particularly since no one else from the project was around. Even though I knew it would be difficult to catch Nim by myself, I decided against enlisting help from students and secretaries. Nim had previously caused enough of a disturbance among the department's secretaries, many of whom wanted to have nothing to do with him.

It took me more than ten minutes to corner Nim at the end of a corridor. During that time he led me on a merry chase through the second- and third-floor corridors and stairwells of Schermerhorn Hall. Even when I had cornered him he was able to keep out of reach by hiding underneath a table at the end of the corridor. Each time I reached under the table, Nim either squirmed out of my reach or tried to bite my hand. Finally, I secured a good grip around his wrist and with just enough of a twist, I was able to persuade him to come out. In the process I incurred my first and only bite from Nim.

I can still recall my anger as I marched him back to the classroom. Nim watched me very carefully as we walked down the hallways, looking for another opening for escape. Unfortunately for him I had my hand wrapped tightly around his wrist, and I was not about to relax that grip until we got back to the classroom. After locking both the inner and outer doors, I sat down with Nim to see what he would do. I hoped that he would sign *sorry* and that he would try to convince me that he intended to behave. Instead he maximized the distance between us and tried unsuccessfully to open the door.

I knew that the bite was a defensive response, but I still felt angry. This was the first time Nim had bitten me and the first time he had tried to run away from me. I felt that unless Nim was made to understand the strength of my anger he might be encouraged to repeat his behavior. After Nim's second attempt to get out of the room, I picked him up and threw him away from me. I was quite surprised by what followed. I had thrown him so hard that he ended up hitting the cinder block wall and not the carpeted floor, as I had intended. I quickly discovered that there was no reason to feel concerned that I might have hurt Nim. Getting up, he half smiled and signed *play*, which I interpreted as a request to throw him against the wall again. I obliged a second time before I realized that, far from punishing Nim, I was engaging him in a game of roughhousing, which he loved.

At that point I was tempted to slap Nim across the face. On other occasions I used that form of punishment after Nim deliberately bit somebody or made a bad mess in the kitchen at Delafield. Nim's reaction to my slap was always one of instant terror. He screamed loudly and spun himself around, often throwing himself into a tantrum. I decided to save hitting Nim for graver infractions. Instead, I tried a new form of punishment. Again I threw Nim against the wall but only to provide myself with an opportunity to make, from Nim's point of view, an unexpected exit from the classroom. Nim's response was instant panic. From the adjacent observation room I could see him banging on the locked door of the classroom. Discovering that he could not open the door, he began to scream an ear-piercing scream. That was followed by a full-fledged temper tantrum. I watched and listened to Nim's temper tantrum as long as I could bear it. As much as it hurt me to see Nim so upset, I knew that the longer I delayed my return to the classroom, the more certain I could feel that Nim understood that I was angry at him.

When I reappeared in the classroom Nim tried to jump into my arms. To underscore his request to be reassured, he signed *sorry* and *hug* many times and continued to whimper. I replied by showing him the bite at the base of my thumb and signing *hurt*. Nim studied my wound and signed *sorry* repeatedly. Then I walked to the outer door and signed *No open* repeatedly. Again Nim signed *sorry* and this time added *hug*. At that point I picked him up and allowed him to hug me.

Of all the forms of discipline I and the other teachers tried, sudden separation and isolation from his teacher proved to be the most effective. Nim's reaction to sudden isolation or loss of attention seemed much like a child's. Since Nim was always within eyesight of his caretaker during his waking hours, it is not hard to imagine the kind of terror and helplessness that he must have felt when he was suddenly left by himself.

I never again had to resort to sudden abandonment as a form of punishment. That is not to say that Nim always behaved when he was with me. In little ways, Nim continued to test me, as he did all of his teachers. But so long as I remained alert and anticipated his attempts at mischief, it was usually possible to restrain him by signing *no* or *bad*. If that type of persuasion did not work, I could stop Nim in his tracks by raising my hand as if to strike him.

Nim often tested me while I drove him to and from Columbia. He had been taught by me and other teachers (though not with the consistency I would have liked) that while he was in a car he was supposed to sit quietly in the front passenger's seat. He was forbidden to jump around or reach for anything in the car, such as the steering wheel, gearshift lever, glove compartment handle, or window knobs. (He was allowed to play with any toys the driver provided for him.) Because going for a drive was such a treat, Nim was usually happy to follow these rules, but there were times, particularly when the car got stuck in traffic, when he became too impatient to sit still. He would express his impatience by various types of mischief: he would look up at the sun visor or at the glove compartment latch in anticipation of reaching for them. Invariably, he would look at me before he tried something. If I responded to his look by signing *no*, Nim refrained from going any further. He often seemed a bit surprised and puzzled by my awareness of his intentions.

On some occasions I could not detect Nim's early warning signals because I was keeping a sharp eye on what was happening outside the car. When I did not respond to Nim's initial probe, he acted as if the coast was clear and tried something more daring. Out of the corner of my eye, I could see his hand reaching cautiously for the glove compartment latch. By signing *no* or by raising my hand, it was easy to discourage Nim from continuing the forbidden course of action.

Disciplining Nim meant more than just controlling him. Discipline was a positive influence in his life in the sense that Nim seemed most content when he was with someone who gently and consistently reminded him what he was and was not supposed to do. I doubt very much that Nim enjoyed creating mischief or attacking people. Often, when he did so, he looked quite troubled. If a dominant caretaker appeared on the scene after he had committed some transgression, Nim would often sign *sorry* or *hug* before the caretaker was even aware of what he had done.

As Nim became older, he expressed his curiosity more directly. Invariably he showed that he was interested in a new person or thing by staring or hooting, but he did not always approach the object of his curiosity directly. Often, his interest was overshadowed by fear. Once he overcame that fear, Nim's behavior ranged from gentle, affectionate exploration to attacks on the focus of his curiosity.

In a strange environment Nim was attentive and alert but often so fearful that he was inhibited about exploring his new surroundings. That, for example, was Nim's reaction to the windowless and stark walls of his Columbia classroom during his first few days there (see Chapter 5, pages 49–50). Even after he appeared to have adjusted to the classroom, Nim reacted strongly to any small changes in it. For example, a camera turret was installed in the wall between the classroom and the adjacent observation room while Nim stayed home, at Stephanie's house, during the Christmas holidays of 1974. I designed it so I could mount still, movie, or video cameras in the observation room and operate them without Nim's seeing who was using them. In order to minimize distortion, the turret was mounted at Nim's height, a few feet above the floor. From inside the classroom, it looked like a Cyclops' eye: a curved piece of unfinished aluminum with a large wide-angled lens protruding slightly from its center.

Nim noticed the turret as soon as he entered the classroom after his Christmas holiday. He hooted and resisted his teacher's efforts to put him down. For most of that day and the next week, Nim's teachers had a hard time distracting him from it. Instead of attending to his teacher, Nim sat at a safe distance from the turret and stared at it for long periods of time. During the next few weeks, he showed less interest in it. But if he was frustrated with an activity or upset with his teacher, he often ran over to the turret and hit it as if to attack the strange creature that was peering at him in his classroom. Then he would scamper back into the protective arms of his teacher with his eyes riveted on it as though he felt that there was a chance that it might strike back at him. He showed a similar but less extreme reaction to the potty when it was introduced to the classroom.

Six months after the turret was installed, a small grant enabled me to replace the shabby, threadbare carpeting in the classroom complex. What used to be a mottled tan carpet had, in one day, become a bright green carpet. No sooner was Nim inside the classroom complex than his hair became erect and he began to hoot. He jumped down from his teacher's arms and, with his hair still erect, began pounding his hands on the floor until he had attacked just about every square foot of the newly carpeted classroom complex.

Nim's fear of new environments often insured that he would stay close to his caretaker. One dramatic example of such "good" behavior was his sitting for his first portrait taken by someone who was not a member of the project. Harry Benson, a free-lance photographer, wanted some photographs of Nim for an article for *New York* magazine. Stephanie and I thought that the best results would be obtained in Stephanie's home. We both felt that Nim would be most cooperative if he were photographed on familiar territory. But Benson wanted to

photograph Nim under the nonportable lights of his studio. With some
trepidation, and making no guarantees about his behavior, Stephanie
and I agreed to bring Nim to Benson. At the time, Nim was a little more
than a year old, old enough to run around but too small to do much
damage.

Inside the studio, Nim seemed awed by the large space, the strange
lights, and the structures that supported them. He noted three or four
new faces: Benson and personnel from the studio. For the portrait, Benson
wanted Nim to sit on a small stool in a three-sided area made of construction paper. About six feet above Nim was a massive light, approximately four by six feet, covered with diffusion tissue. Our problem was to
get Nim to sit under powerful lights in a totally strange environment
long enough for Benson to get the picture he wanted. Stephanie and I
took turns sitting next to Nim just out of view of the camera. Each
time Nim tried to get off the stool, one of us would sign *sit, stay, Nim,*
or whatever else we could think of to get him to stay put. At roughly
fifteen-minute intervals, we allowed Nim to take short breaks, during
which we encouraged him to play with some of the nonbreakable equipment at the studio.

Nim exceeded our wildest hopes about cooperating for his portrait.
Benson took many excellent photographs, one of which appeared on the
cover of *New York* magazine in February 1975. All in all, Nim spent
close to two hours at the studio. During most of that time he sat quietly
on a stool. In hindsight, it is easy to recognize the factors that contributed to so much unexpected cooperation. Nim was fascinated by the
strange environment, strange equipment, and many new people doing
strange things: moving lights, turning them on and off, operating
cameras, changing lenses and film, and performing other photographic
chores.

A few months later, we brought Nim to a television studio, where
he, Stephanie, and I were taped for a talk show. There again he was
quite docile and cooperative. Undoubtedly, the powerful lights, the TV
cameras, and the difficulty of looking past the lights to see who was
operating the cameras contributed to Nim's passive behavior and his
eagerness to stay close to Stephanie or to me. Nim's next visit to a TV
studio caused us more concern. On that occasion, Stephanie and I brought
Nim for a taping of the "David Susskind Show," and initially Nim was
just as cooperative as during his first TV taping. But about halfway
through, he began to get restless. He signed *hug* to Susskind, who was
at first too stunned to respond. After I assured him that it was all right to
hold Nim, Susskind gingerly took him in his arms. A few moments later
Nim began grooming Susskind's face. Nim must have sensed that Susskind
did not feel terribly comfortable and that it would be easy at this point
to get away. He signed *down* several times. For me, that was a warning

that Nim was about to take off. I signed *hug* and got him to sit in my lap, but for the remainder of the taping, I had my hands full restraining a chimpanzee who was more than ready to explore his strange surroundings.

Nim's curiosity about his environment manifested itself even in familiar settings. Within a few months of his move to Delafield, he seemed content both with his new home and with his caretakers, Laura, Amy, Andrea, and Walter. One warm evening late in October, Walter was reading contentedly in his room and listening to classical music. Absorbed in his book and the music, Walter was unprepared for the sudden entrance, through one of his windows, of a terrified chimpanzee. What we thought was a secure bedroom had an insecure window. But how could we have guessed that a twenty-month old chimpanzee would open a third-floor window and jump down onto a pitched roof more than six feet below? Equally surprising and terrifying to Nim's caretakers was the eight-foot jump that he had to make in order to get from the roof under his window to the roof over Walter's bedroom. I have no way of knowing how long Nim contemplated jumping from his window to the roof below or what he thought he might discover. On that particular night, however, his curiosity about what existed outside his window (and perhaps also about the sounds of Laura's and Amy's voices emanating from the back lawn) was stronger than his fear of jumping down onto the gabled roofs of the Delafield mansion.

Almost two years later, Nim again escaped from his bedroom under somewhat different circumstances. During the day before his escape, he had been visited by a film crew who put him through a demanding schedule in order to obtain footage showing what he did during a "typical" day. Nim's docility and cooperation lulled us into thinking that everything was fine. He showed no resistance to the many requests that he sit longer at the breakfast table, that he repeat various activities with his teachers, and so on.

Footage of Nim sleeping and getting up was of special interest to the camera crew and to me. He had not been previously photographed in his bedroom because even if one had been able to tiptoe undetected to the door of his bedroom, it would have been impossible to open the door and set up a camera without waking him. In order to minimize the intrusion of a camera in Nim's bedroom, I removed a section of the wall across from his bed and modified it so that it could be reinserted quietly and securely. My plan was to use the opening as a portal for a camera lens. On the day before we were to film Nim's slumber and awakening, the movie crew set up a tripod and camera just behind the opening I built for the camera lens. A few moments before the filming was to begin, I gingerly removed the section of wall in front of the lens and draped some black velvet cloth around the lens. These goings-on woke Nim, but

after he sat up and stared at the lens, he pulled his blanket over himself
and went back to sleep.

What I thought would be a difficult photographic challenge yielded
clear pictures of Nim getting up and greeting Joyce Butler as she entered
his bedroom. Nim's docile mood continued when he was taken to the
bathroom. There I was able to photograph him using the toilet and brushing his teeth. A few moments later he completed his morning ritual downstairs by dressing himself under the watchful eye of Bill Tynan. So
innocuous was Nim's reaction to my photographic intrusion, that I
decided to film Nim sleeping and getting up the next morning. In anticipation of that filming, we left the camera in place, with the lens protruding
through the drapes that covered the hole in the bedroom wall.

That evening, Bill Tynan called me at home to tell me that Nim had
escaped. Apparently Nim had been more concerned about the events of
the day than we thought. He had found a way to climb up to the small
hole for the lens, push the rigidly placed camera out of the way, squirm
through the opening, and climb out the open window of his playroom. He
then lowered himself to the ground by clinging to the temporary power
lines rigged to the playroom window for the camera lights. Some time
later, Nim climbed up to the terrace outside Joyce Butler's window.
Without any warning, Joyce was visited by an unannounced guest who
had to satisfy his curiosity about the strange events he experienced while
being filmed.

Nim "escaped" from his bedroom or from some other part of Delafield about a dozen times during the two years in which he lived there.
Not all of these escapes were expressions of curiosity. The two I have
described were exceptional in that they seemed motivated mostly by
Nim's desire to explore an unknown feature of his environment. The other
escapes expressed his sadness over the departure of a caretaker he was
fond of or his frustration when he was denied some regular feature of
his life.

All four of Nim's original caretakers at Delafield left the project
during or immediately after the summer of 1976. Nim reacted in different
ways to each departure, but always with some kind of agitated behavior.
After Andrea Liebert left, Nim refused to work on any tasks in which
he was required to name or to sort colors—tasks that had been introduced to the project by Andrea—even when asked to do so by teachers
with whom he was normally cooperative. He was quite destructive as
well. Often he would tear the colored papers put down before him or
bite the crayons he was given or throw them against the wall. After

Nim signs *finish*.

Walter left, Nim threw temper tantrums on Sunday mornings. For more than a year, Walter and Nim had gone through an elaborate ritual every Sunday morning of laying out utensils and ingredients in the kitchen in preparation for making pancakes and syrup, a dish that always delighted Nim. Laura and Amy told me that they were always reminded that it was Sunday morning when they heard Nim's hoots of delight reverberate throughout the Delafield mansion. Laura and Amy took over Walter's Sunday morning session after he left, but no matter what they did, Nim threw temper tantrums, sulked, and stared at the kitchen cabinet in which Walter kept the utensils and ingredients he used for making pancakes.

About a month before Walter planned to leave, Laura, Walter, and I discussed how best to prepare Nim for the loss of the many activities he had enjoyed with Walter. One thing Nim especially loved was to be taken on walks through the streets of Riverdale. He was fascinated by the people he met, particularly children. During these walks, Walter watched Nim carefully. At the slightest sign that Nim might hurt someone, Walter interceded either by warning him or by picking him up and walking away. Thanks to Walter's vigilance and dominance, Nim was exceptionally well behaved on these walks.

Nim responded very tenderly to children. I suspect that this was due to their uninhibited openness and to their interest in playing with him— an interest whose intensity was unequaled by most of his adult caretakers. But it was rare for Nim to have the opportunity to dominate another creature, and with a caretaker who was less in control than Walter, he would sometimes try to dominate a child by hitting or biting. When Walter was around, Nim refrained from attacking children in any way and instead absorbed himself completely in the various games that he and his new-found playmates invented. He became so attached to a few of the neighborhood children that he could recognize them from afar. When he did, he often turned to Walter and signed *play*, *hurry*, or *run*.

Walter, Laura, and I decided that until we found someone to replace Walter, it would be wise to phase out Walter's walks with Nim. As an alternative, Walter offered before his departure to engage Nim in other activities on the spacious acreage of the Delafield estate. At first Nim seemed to tolerate not going on his accustomed walks. He often led Walter to the gate of the property, where he signed *out* or *play*. But when Walter distracted Nim by pulling out a large ball or by running away from him, Nim seemed to forget about his interest in leaving the estate.

During one sunny afternoon when Nim expressed interest in going for a walk, Walter tried his usual diversionary tactics. Just as he was about to distract Nim from the gate, one of Nim's neighborhood playmates shouted to him. In a flash, Nim bolted away from Walter and out

Walter and Nim playing on the grounds of Delafield

through the gate. For the first time, Nim was outside the Delafield property by himself. He looked around for the child, who had apparently run
off when she saw Nim racing toward her, and then he loped into the
backyard of the house closest to the edge of the estate. Nim had often
met the residents of that house, a psychiatrist and his family, but had
never been invited inside. This time the open front door and the sound
of voices coming from within the house were all the invitation that Nim
needed. A few minutes later, Walter, quite out of breath, was trying to
calm a startled family whose afternoon had been suddenly disrupted
by the unannounced visit of their chimpanzee neighbor.

After this escape, it was decided that *all* of Nim's caretakers would
be required to keep him on a lead. The issue of the lead had a long and
checkered history. When Nim was only six months old, Stephanie kept
him on a lead in her house whenever she wanted to be free from the
ordeal of watching everything he did. A similar practice is followed by
many European mothers, who put a harness on their babies and then
attach a long lead to the harness. That arrangement allows the baby
freedom to wander around a substantial area while preventing him from
running away. As Nim got bigger, we outfitted him with a variety of
harnesses and leads.

New teachers were always asked to keep Nim on a lead, particularly
because of the catch-me-if-you-can games he played with them both

inside and outside the mansion. By using a lead a new teacher could get off to a good start in establishing dominance as well as save experienced teachers from being called in to catch Nim when the new teacher was exhausted by Nim's cat-and-mouse maneuvers. Experienced teachers, including myself, resisted keeping Nim on a lead because we wanted to show that we could dominate Nim without recourse to external aids. We all prided ourselves on getting Nim to behave simply by signing to him or by making a threatening gesture.

Nim's escape from Walter showed that even a dominant caretaker could not be sure that Nim would always obey his commands. Nim's proven ability to escape from the house or from a caretaker was obviously quite dangerous: he might get hurt if he ran off by himself; or he might hurt someone who was trying to catch him, particularly someone who was inexperienced in handling a chimpanzee; or he might cause damage to other people's property. The consequences in any of these cases could be tragic. Apart from the danger of physical injury to Nim, fear of his injuring someone or destroying property might create pressure to have him moved out of Delafield.

Even with the lead, Nim succeeded on a few occasions in escaping from experienced caretakers. Nim was quite crafty in lulling his caretaker into thinking that he was totally preoccupied. Once Nim perceived that the caretaker was not holding on to the other end of the lead, he raced off as quickly as he could. But it was much easier to catch Nim when he was wearing the lead than when he wasn't. All the caretaker had to do was to grab the trailing lead or step on it.

After Laura left, Nim escaped twice within the same week, both times from experienced caretakers while wearing his lead (Amy Schachter on one occasion, Susan Quinby on the other). Both escapes occurred during the early evening, shortly before he was to be put to bed. Prior to one escape, Susan was sitting with Nim curled up on her lap. He seemed almost asleep. Suddenly a child's voice from the distance aroused him. In an instant, his reverie with Susan was broken. It took three caretakers to corner Nim and a howling dog to make him upset enough to come back to Susan.

Even though this escape seems to have been triggered by a child's voice, I often felt that Nim's interest in leaving the Delafield property, an interest he expressed without necessarily trying to run away, was motivated by a desire to find the caretakers who had abandoned him within the span of two months. Particularly after Laura left, Nim seemed depressed and inconsolable. New caretakers had trouble putting him to bed. At first, Nim would not let them leave his bedroom. Once they did, he would often cry piercing screams for hours on end.

The revolving-door manner in which caretakers cycled through Project Nim brought out the dark side of Nim's personality. If it had been possible to have him looked after by the same group of caretakers, I am sure that his amiable and affectionate nature, which was apparent to anyone who got to know him, would have expressed itself more often. As it was, Nim was too often frustrated by the sudden disappearance of a caretaker he had grown to love. New caretakers sometimes had a hard time believing that Nim could behave as docilely as I and other experienced teachers had claimed he would.

An ideal environment for Nim, both from his and from his teachers' points of view, was one in which he could satisfy his curiosity about the world in ways that were not destructive. Indeed, a good teacher could be defined as one with the imagination and the wherewithal to create such an environment. The reward for such efforts was access to what was, from my point of view, the most fascinating aspect of Nim's personality: his desire to explore, to manipulate, and to communicate about his environment.

Only a small fraction of Nim's teachers and caretakers got to know his inquisitive and pensive side. Most teachers barely developed enough control over him to insure that he didn't act like a brat. Most of their time was spent feeding him and playing with him simply because they did not have the control needed to direct his attention to such challenging activities as looking at pictures, drawing, or playing with toys. As described in the next chapter, these and other activities were used initially to exploit Nim's curiosity. In time, they were used to teach him the names of various actions, relationships, objects, and their attributes.

During those sessions in which Nim was well behaved and completely attentive I felt that he regarded me as some sort of magician and that his attentiveness to me derived in part from his belief that I had special powers to make colors with crayons, pull pictures of animals and flowers out of my pocket, blow up balloons, open boxes that were carefully secured, and so on. I believe that part of this power derived from my general dominance over Nim. If I could sense when he was about to do something, particularly something mischievous, it was reasonable for Nim to conclude that I could perform other magic as well.

Nim's personality resembled a child's most closely in situations in which he respected his teacher and regarded him or her both as someone to please and someone who could stimulate him. Even without sign language, that kind of relationship was rich in communication about Nim's and the teacher's feelings as well as about external objects of mutual interest. All of Nim's teachers who built up that kind of relationship experienced the joy of give and take with a humanlike personality. They could also see how essential that personal relationship was to Nim's taking the important step to a higher level of communication: communication through sign language.

8

Nim's Teachers

Many human infants have been raised exclusively by a single parent. Though I know of no example of a single person taking full responsibility for raising an infant chimpanzee, I see no reason to doubt that it could be done. What does seem impossible is for a single person to socialize an infant chimpanzee and at the same time teach it to sign and collect scientifically useful data about what the chimpanzee signed. The strength and agility of a growing chimpanzee make the task of rearing it exhausting enough. It would be too draining to add to that task the intense effort needed to amplify and focus the chimpanzee's less than human intelligence in ways conducive to learning sign language *and* also keep records of the chimpanzee's signs and the conditions under which they occurred. That kind of work could be sustained only for short periods of time.

My original plan for teaching Nim to communicate via sign language was to rely on a small group of gifted teachers of sign language who could commit themselves to work with Nim for at least five years. By limiting the size of this group to six to eight teachers, I hoped to establish strong social bonds between Nim and his teachers and avoid the emotional disruptions caused by replacing one teacher with another. By breaking up each day into three or four periods and by assigning responsibility for each period to a different teacher, it should be possible for all teachers to apply themselves fully in their interactions with Nim.

The limited resources of Project Nim prevented me from putting this ideal arrangement into practice. During the forty-six months Nim was in New York, he was taught by no less than sixty teachers. Aside from their fascination with Nim, his teachers had little in common. Their ages ranged from nine to fifty-three. Many of them were students in high school, college, or graduate school. Others were housewives, social workers, speech pathologists, or teachers of normal and disturbed children. Some spent their nonproject time in such occupations as clothing manufacture and personnel management, areas quite unrelated to the business of teaching sign language to a chimpanzee.

This chapter is about some of the unusual people who were attracted to Project Nim by the opportunity to teach sign language to a chimpanzee, how they learned to work with Nim, how their individual teaching styles fostered Nim's linguistic development, and how they exerted themselves beyond my most extravagant expectations. The backgrounds of Nim's teachers and how they were recruited are described in Appendix B.

Those people selected as teachers in training were given three immediate assignments: to observe Nim in his Columbia classroom, to read revelant background material, and to study sign language intensively. Prospective teachers got their first glimpse of Nim by watching classroom sessions from the observation room. Laura, I, or another experienced teacher usually described what was happening. Each candidate had to be told the many basic details of establishing a good teaching rapport: why Nim's teachers worked with him on the floor instead of towering over him from a standing position, how to pace an activity, the importance of eye contact, what to try in case eye contact was lost, how to anticipate Nim's reaction to a new activity, how to capitalize on it, when Nim was and was not allowed to drift away from his activities, and so on.

It often proved difficult to translate my own fairly clear idea of how Nim's teachers should work with him into practice. Just as the same philosophy of raising a child is translated into as many styles as there are parents involved, different aspects of my philosophy of working with Nim were expressed through the different personalities of his extended family of teachers. As will become clear from the examples that follow, my advice had relatively little effect on how a teacher and Nim interacted, especially during their first encounters. I ended up feeling that my main contribution to the success of a teacher's work was the selection of that teacher, a choice based largely on my feelings about how the teacher and Nim were likely to interact. Once a new teacher was introduced to Nim, the advice he or she received from me or from other teachers was overshadowed by the chemistry that developed between Nim and his new human companion.

As is true among humans, Nim's first encounter with a new teacher often proved decisive. In less than a minute, the kind of relationship that would develop between Nim and the new teacher was often all too clearly apparent. New teachers were always introduced in the presence of someone Nim trusted. This undoubtedly made Nim feel safe in approaching the stranger. He was rarely impassive. Even though I could not ask Nim what determined his initial reaction to a new teacher, it seemed clear that the teacher's attitude was crucial. Nim's behavior suggested that he was an excellent reader of body language and facial expression. Often it seemed that he could see through people's attempts to disguise their negative feelings. In her diary notes, Laura described an important aspect of a successful relationship with him: "It was more than just being relaxed. Nim had an uncanny ability to read one's feelings. I always felt that I had to be honest because he understood me . . . he made me feel 'naked.' "

On a number of occasions when I had felt sure that Nim would react positively to a new teacher, he instead acted very belligerently. Later it would become clear that the new teacher had not been as relaxed or as open to Nim as I had surmised from his or her verbal behavior and confident airs. If a new teacher was truly relaxed and as curious about Nim as Nim was about the teacher, Nim's reaction was quite positive. Most often he gave his new teacher an uninhibited chimp hug. On other occasions he would rush over, sit down next to his new teacher, and begin to explore and groom the teacher's face and hair. Women with long hair and men with beards or mustaches seemed especially attractive. But physical characteristics could not compensate for negative feelings a teacher might have toward Nim. Nim had a very simple view of how you accepted him: all or nothing.

Almost never did a new teacher win a first confrontation with Nim. To do so, the teacher would have had to discipline Nim for any bad behavior. Knowing when and how to discipline Nim took quite a bit of experience. Particularly important was one's timing and confidence in administering discipline. Nim's response to the conventional discipline that one might mete out to a child was to escalate and become more aggressive. The problem of disciplining Nim was even worse when an experienced and a new teacher were present at the same time. Nim's aggressive reaction to discipline from a new person was often aggravated by the presence of an experienced teacher. Nim acted as if he assumed that the experienced teacher would protect him from any serious counter-attacks initiated by the new teacher.

Once a teacher got off to a bad start with Nim, it was very difficult to reverse things. If Nim attacked a new teacher during their first encounter, he would often strike out at that teacher with even less hesitation the next time they met. In many instances it seemed as if Nim was taking advantage of the rare treat of being with someone less dominant than himself. As Nim got older the likelihood that a new teacher could reverse a bad start became less. Even if the two could effect a temporary truce, it usually didn't last long enough to redirect the relationship. There was little that Laura, I, or other teachers who observed a bad initial encounter could do. We could shout encouragement and suggestions as to how the new teacher might win a fight with Nim. In the heat of battle, however, the novice found it difficult to attend both to Nim and to advice offered from the observation room. It only made matters worse if an old hand stepped in to try to calm Nim. Even if the experienced teacher succeeded in getting Nim to refrain from attacking the new one, Nim ended up regarding the new teacher as weak, and the next time the two were alone together, Nim would resume his attacks with increased vigor. Once it became clear that a new teacher could not withstand Nim's testing it was pointless to continue. For that reason many otherwise qualified teachers had to give up working with Nim.

Each loss of a new teacher was costly and frustrating to Nim, to the teacher concerned, and to everyone else on the project. Nim became unnecessarily aroused and after fighting with a new teacher was more difficult for his regular teachers to handle. The unsuccessful teacher had little to show for his or her time except the experience of what it was like to fight with a young but far from helpless chimpanzee. Laura, I, and others who worked hard to train new teachers had to start the exhausting process of locating and training all over again.

Once Nim's interest and respect were sufficient for him to behave well, the new teacher was faced with the greater challenge of getting Nim to sign. The teacher had to learn how to interest Nim in various activities that would help him learn the preverbal concepts that he might later be motivated to sign about. This step was the most intuitive part of working with Nim. New teachers could be taught to perform standard chores such as feeding, toilet training, and putting Nim to sleep without much difficulty. But how to instruct Nim was a skill that could only be taught by example, if indeed it could be taught at all. By and large it was a skill that new teachers had to develop by themselves.

In order to stimulate new teachers to develop a teaching style with which they felt comfortable, we encouraged them to observe experienced teachers working with Nim and to watch videotapes of sessions in the classroom. Laura and I tried to point out many of the nuances of Nim's behavior. Nim was shown at his best and at his worst: when he attended diligently to his teacher, when he was bored and simply stared at the walls of his classroom, and when he got up, walked away from his teacher, and began to act like a brat.

Andrea Liebert, a twenty-one-year-old graduate of SUNY college in Buffalo, New York, was a good example of a new teacher who studied Nim and how he was taught by other teachers before she began to work with him. Before Andrea began to work on the project (in November 1975), I knew that she would be available for only nine months; she was in transition between college and a graduate program in speech pathology at Peabody College, in Tennessee. But even for nine months, I thought that Andrea's strong background in American Sign Language and her unabashed interest in teaching Nim to sign made her a good bet as an effective teacher.

Andrea watched sessions with Nim almost on a daily basis, most of them taught by Laura and containing many excellent examples of good teaching technique. Other sessions, taught by teachers of lesser ability, often provided nothing more than examples of a bored chimpanzee who eventually created much mischief in the classroom. Andrea learned from both kinds of examples. After weeks of careful observation of Nim, she herself began to teach him to sign. Within three months she was one of Nim's best teachers.

Even though Andrea relied mainly on Laura as a role model, there were vast differences in their teaching styles. Laura liked to overwhelm Nim with spontaneous surprise and, when appropriate, punishment. She shifted easily between expressions of joy, anger, boredom, and praise in working with Nim. Andrea's attitude was more sedate and measured. She was more detached than Laura and only rarely responded to Nim with strong emotion.

In order not to tower over Nim and to establish the strongest possible contact with him, all teachers sat with him on the carpeted floor of the classroom. Andrea and Laura, however, assumed different physical positions in working with him. I suspect that this was due to differences in their confidence about controlling Nim and getting him to pay attention to the task on hand. Andrea held Nim close. Often Nim sat on her lap with one arm curled around her neck. Laura, on the other hand, sought to maintain a small distance between herself and Nim, as if to make Nim less dependent on her during his lessons. When Nim performed well, however, Laura often hugged him abruptly or engaged in a brief game of play-tickle.

In different ways Laura and Andrea were consistent in what they demanded of Nim. Good behavior was rewarded and bad behavior was punished, albeit with different intensities. From Nim's point of view, however, it was clear what was considered good and what was considered bad. This kind of consistency was an essential aspect of working with Nim that often proved difficult to transmit from teacher to teacher.

One important aspect of any teacher's success with Nim was satisfying his developing sense of "justice." Nim expected praise for good behavior, for example, when he signed correctly about pictures he was shown or when he solved a puzzle. He also accepted punishment from a teacher if it was for something he understood to be bad, like breaking a toy, mouthing a doll, or running around the classroom. But he could be quite uncooperative if one teacher punished him for something that other teachers had accepted or if one ignored something that the others praised.

Some of Nim's new teachers began to work quite effectively with a minimum of observation. Susan Quinby, whose self-assurance enabled her to take charge of almost any situation, had no hesitation in taking on Nim after hearing my brief spiel about the project at a group interview. Susan was recruited from Professor Lois Bloom's class of developmental psycholinguistics at Teachers College. When she applied to work on the project she was also an experienced teacher of children with learning disabilities.

After observing only a few sessions, Susan went to work by herself. From the start, the relationship between Susan and Nim was quite positive. Susan amplified Nim's interest in her by bringing highly stimulating activities to her sessions, inspired both by what she saw and heard of other teachers' work with Nim and by her teaching experience with children.

A new way of painting that Susan taught Nim is a good example of her talent for creative improvisation. Painting on a pad of paper with magic markers or paint brushes was already one of Nim's favorite activities. One of the first things Susan brought to the classroom was a new set of painting materials. Instead of paper, Susan used shiny aluminum foil. First she introduced Nim to the aluminum foil, a material that fascinated him because of its shininess and the faint image of his reflection it produced. She then showed him how to handle the foil without tearing it. Soon he was helping her to paste the foil to the wall. In order to increase Nim's curiosity, Susan stretched each step out as long as possible. After Nim had banged his hands a number of times on the aluminum foil hanging on the wall (his way of saying he liked it), Susan squirted the contents of a can of shaving cream into two small bowls. Finally she added red and blue vegetable dyes to the bowls. Nim had before him a new set of "paints."

Predictably, Nim was fascinated by the whole process. With one arm around Susan's neck he touched the colored shaving cream and then smelled it and tasted it. Deciding it was harmless, he briefly immersed his hands in the mixture and began to play with it much as a child would play with mud or some other gooey substance. When Nim's excitement began to ebb, Susan rubbed some red cream onto the foil, and then some blue. Nim caught on quickly and began to apply the colored creams himself. Before long he was rubbing the cream on the foil with both hands, making interesting three-dimensional constructions of different colors. All in all, this activity took about forty minutes, quite a long time for a two-year-old infant chimpanzee with a short attention span.

Painting and other activities used by Nim's teachers served as a basis for topics of conversation in sign language. A few months after Susan taught Nim how to paint on aluminum foil with shaving cream, Nim was making the signs for *paper, red, black, blue,* and other colors. Susan can not of course be given exclusive credit for teaching these signs. They were also emphasized in activities prepared by other teachers, for example, cleaning up with a paper towel or working with colored paper (identified as *red, black, blue,* and so on) in tasks that required Nim to place pieces of paper of the same color in the same pile. But in this and in many other instances, Susan's ideas contributed substantially to Nim's interest in working in the classroom.

Having noticed that Nim was interested in pictures of faces, Susan asked him to help her paste a picture of a face on the classroom wall. Earlier Susan had noticed that Nim had paid special attention to this face while thumbing through a magazine. She then cut out the eyes, the nose, and the mouth of an identical face and asked Nim to paste each face part on the appropriate area of the face that he helped to attach to the wall. After Nim mastered this task, the eyes, ears, nose, and mouth of the face on the wall were cut out, and he was asked to paste these on as replacements for the missing parts from the face on the wall. As a result of this activity, as well as pointing to face parts of various dolls with which he played, Nim later learned the signs *eye, ear, nose*, and *mouth*.

We taught Nim the sign *cat* by showing him pictures of cats in books. Once he had learned to sign *cat*, Susan brought her own cat to the classroom in a small traveling case. Nim sensed immediately that there was a living creature in the box. When he discovered that he could not open it, he signed *open* repeatedly, but Susan was in no rush to comply. First she signed to Nim *Who in box?* Nim didn't seem to understand and kept signing *open, me open, Nim open*, and *hug* (as if to say I want to hug what is in the box). After Susan signed to Nim, *Cat in box*, Nim began to sign *cat, cat hug, cat me* at a furious rate. Finally Susan opened the box and took out the cat. The sight of Susan holding the cat further intensified Nim's signing. Now he began to sign combinations such as *hug cat, me hug cat, me hug*, and *me cat*. When Susan finally allowed Nim to hug and play with the cat, he grinned broadly in an unmistakable smile. Surprisingly, the cat seemed quite tolerant of Nim's curiosity and antics even though Susan had to restrain Nim occasionally from pulling too hard on the cat's tail or leg.

Nim sometimes seemed jealous of the cat. If Susan held on to the cat for a long period of time, Nim began to act in a rowdy way and did whatever he could to call attention to himself, like climbing into the cat box himself. Once Nim sat down in front of Susan as she was about to feed the cat some yogurt. Nim enjoyed pretending to feed his dolls and puppets. He seemed to understand that there was no danger of an inanimate object ingesting food in which he was interested. When Susan offered a spoon of yogurt to the cat Nim took the spoon out of her hand. He then tried to feed the cat himself, just as he had done while playing with a doll or a puppet. To his amazement, the cat licked the spoon clean. Nim was not so generous the next time. When Susan signed to Nim that he should feed the cat, he deliberately offered the cat an empty spoon. He then filled the spoon with a generous helping of yogurt and fed himself.

Despite Susan's many innovations in the classroom, I had one reservation about her teaching style, a reservation I often discussed with

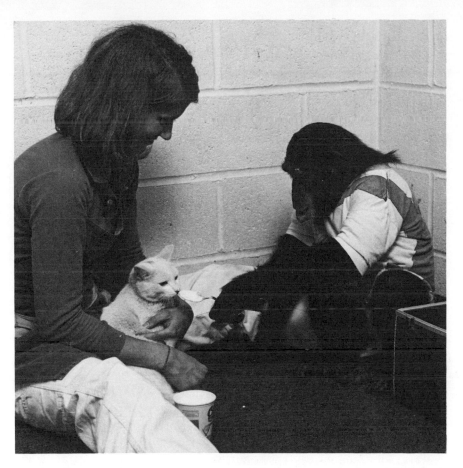

Feeding Susan's cat

her. Even in situations as highly motivating as Nim's playing with the cat, I felt that Susan drilled Nim with too many repetitions of the same question. For example, after Susan taught Nim her name sign (the right hand, with a closed fist, crossed over to touch the left arm just below the left shoulder), she would repeatedly ask Nim, *Who me?* before shifting to another question or activity. Nim eventually responded, but as far as I could tell he became quite bored by this activity and responded to Susan's questions more because of her persistence than because of an interest in having a spontaneous conversation with her. I could see the same problem occurring when Susan brought her cat to the classroom. I didn't see what was to be gained by asking *Who that?* or *Who hug?* five times before Nim was allowed to hug the cat. Indeed I often worried that the repetitive use of a sign could diminish its meaning for him.

Other teachers asked Nim about the things they wanted him to sign about, but it struck me that they did so more spontaneously, often shifting abruptly, conversationally, from one kind of question or activity to another. I tried to make this clear to Susan by having her watch with me from the observation room or on videotape what other teachers were doing. Susan accepted my observations and criticisms of her teaching style quite graciously but said she thought Nim needed to be drilled if he was going to sign properly and accurately. She also felt that Nim had to be dealt with pretty much as one might deal with a child who had a severe learning disability. That meant not going on to the next part of the lesson until the current problem had been mastered.

I tried to convince Susan that the best approach to Nim was to regard him as a young chimpanzee and not as a child. To reinforce this philosophy I reminded her of how well Laura and Andrea worked with Nim. Susan respected Laura and Andrea, both of whom felt strongly that Nim should not be regarded as a child. But Susan believed just as strongly that the techniques she had learned at Teachers College for teaching children with learning disabilities were especially applicable to Nim. That background, and her success in teaching Nim, convinced her that her approach would ultimately lead to significant gains in Nim's linguistic development.

With two of Nim's other outstanding teachers I had disagreements not about *how* but about *where* he should be taught. Amy Schachter and Walter Benesch, who lived at Delafield, preferred to work with Nim at home rather than in his classroom. When they tried to work with Nim at "school," they found it difficult to adapt to the many activities I suggested they try. Nim himself may have contributed to this problem by resisting Amy and Walter in their new role of classroom teacher.

For a variety of reasons I resisted excusing Amy and Walter from sessions in the classroom as long as I could. I felt that it was important that each teacher observe all of Nim's other teachers. This would provide a concrete basis for sharing experiences with Nim and would result in a greater consistency of teaching methods. And I wanted to videotape all of Nim's teachers in a standardized setting. Videotaping with a portable system was feasible at the house, but it could not be done without Nim's being distracted by seeing the cameraman. In the classroom, a camera could be positioned innocuously in the wall so that Nim was unaware of the operator. Finally, I felt that the activities that Nim's teachers worked on in the classroom would be reinforced strongly if Walter and Amy, two people whom Nim liked very much, also worked on these activities.

When Amy and Walter told me that they didn't want to work at Columbia, I suspected at first that they were trying to avoid being observed. Observation was difficult at Delafield because once Nim noticed an observer, he wanted to greet and play with that person. On some occasions, however, I was able to observe teachers like Walter and Amy from behind a tree or through a window without Nim's being aware of my presence. In this manner I slowly learned how effective Walter and Amy were with Nim at Delafield. Although they had difficulty in adjusting to the classroom, both Amy and Walter thrived on the relaxed and unstructured opportunities for signing that arose at Delafield. As described in the previous chapter, Walter went through an elaborate Sunday-morning ritual of preparing a pancake breakfast for Nim. At other times he took Nim for walks on the back streets of Riverdale or romped with him through the grounds of the estate. Whenever possible he attached signs to these activities in a conversational manner. As far as I could tell from watching Walter and reading his reports, he was just as diligent in requiring Nim to sign at home as were the best teachers in the classroom.

Amy's style with Nim was similar to Walter's, though much quieter. Instead of the physical activities that Walter and Nim shared indefatigably, Amy encouraged Nim to help her prepare meals, do the laundry, and clean up around the kitchen. Once, Amy reported that Nim even tried to pull a needle through a sock she was darning. In such situations there was an abundance of two-way communication between Nim and Amy. I was struck by the very strong bond that developed between them and the richness of Amy's reports about Nim's overall development. My impressions were confirmed by Laura, who served as an invaluable liaison between the classroom and Delafield. Laura not only read all of the reports of Nim's Delafield caretakers but also spent a lot of time observing their work and taking part in "family" activities.

Though my problem with Amy and Walter was different from my problem with Susan, the rationale for solving both was the same. In the final analysis, Nim was only as content as his teacher. After a certain point I sensed that if I pressured teachers to deviate too much from the styles with which they felt most comfortable, their effectiveness would decrease markedly. Better to have an effective and enthusiastic group of teachers whose styles and predilections did not overlap completely than to create irreversible tensions that would eventually affect Nim.

Teachers like Amy, Walter, Laura, Susan, and Andrea, along with many others who replaced them, carried the main load of working with Nim, but there were other teachers who contributed significantly to the project, albeit on a sporadic basis. Some of them were quite young. Often they lacked a full understanding of the intellectual questions that the project sought to answer. Nevertheless, they supplied a spark that often seemed absent in Nim's older teachers. A good example is Fred Bever, Tom Bever's twelve-year-old son. Fred brought a Hula-hoop for his first meeting with Nim. It was love at first sight for both Fred and Nim. Nim screamed with joy and signed *play*. Nim also squealed with delight when other young teachers such as Maggie Jakobson and Jennie Lee came to work with him in the classroom. What Nim's younger teachers lacked in rigor they more than compensated for in spirit. Indeed I often sensed that Nim found the sessions with his younger teachers a relief from the rigorous pedagogy to which he was regularly subjected.

Nim also accepted and seemed to welcome the slower pace of such older teachers as Jean Baruch and Dorothy Moscow, whose sessions were neither as playful as those run by Fred or Maggie nor as didactic as those run by Laura and her contemporaries. They nevertheless seemed just as stimulating. There I was impressed by the mellowness of their style and by the way they got Nim to sign with a minimum of fuss.

The primary source of information regarding what Nim signed was the teacher's detailed report written after each session. These reports listed every sign and sign combination that Nim made, with a brief description of the context in which the signing occurred. To produce their reports, Nim's teachers played back miniature tape cassettes that they had dictated, in a whisper, while working with Nim. As Nim's signing ability increased, so did the contents of his teachers' reports. This in turn increased the time needed to tabulate the contents of each report and to reconcile it with the contents of other teachers' reports. The final product was a more detailed corpus of utterances than had been obtained from any other chimpanzee and, as far as I know, from any child. Laura and I read each report with an eye to improving consistency and identifying topics that we felt needed discussion at our weekly meetings. Laura also used the reports as a basis for writing weekly summaries that described Nim's linguistic and social development.

With a few exceptions—Laura, Bill Tynan, Dick Sanders, Joyce Butler, Bob Johnson, and myself—the people who worked on data analysis did not teach Nim. This was just as well because the demands of each type of activity seemed to grow geometrically. It wasn't long before it became impossible to do both tasks well. Yet it would probably have benefited the teachers to see, in the form of raw data, the general trends that were emerging in Nim's signing. It would also have benefited the data analysts to see how the data on which they worked was obtained. At times the data analysts were able to get a small glimpse of Nim's signing by watching a classroom session through the one-way window, but that kind of vicarious experience was no substitute for having a conversation with Nim and seeing, from a teacher's point of view, how Nim mastered yet another aspect of sign language.

Bill Tynan combined the jobs of teacher and data analyst best. When he joined the project in September 1974, Nim was ten months old. Bill was a full-time undergraduate in Columbia's School of General Studies. He continued to work on the project as a full-time graduate student in Columbia University's department of psychology. From the start, Bill

With Jean Baruch

volunteered his services whenever he found a chance, which seemed to be just about all the time. Right up to the sad moment when Nim was returned to Oklahoma, I could always count on Bill to handle any of the complex problems that the project generated regularly. In addition to working with Nim for the usual two to three sessions a week, Bill assumed responsibility for coordinating a group of ten volunteer data analysts who did nothing but tabulate the frequencies of Nim's many signs and sign combinations.

Data analysis became particularly frenzied prior to deadlines for grant applications. Though I didn't like working around the clock in order to meet a deadline, my reason for doing so was quite clear: we needed the money. But I could never quite figure out what motivated Bill and the other volunteers who often worked well into the night, seven days a week, in order to complete laborious compilations of Nim's signs. I often felt that the volunteers would have resented my asking them to do what they did. All I had to do to motivate them to do awesome amounts of work, with such gratifying displays of conscientiousness, was point out why certain analyses were needed and what they meant. I can never thank those volunteers enough for the sheer quality and the quantity of the work they performed.

During the first few years of the project most of Nim's teachers were female, due more to practical considerations than to Nim's or my predilection for females. Women generally had more time to contribute than men. Another factor was the unmistakable maternal response of many females, particularly when Nim was a cuddly infant. As Nim got older and stronger, the number of male teachers increased steadily. In September 1976 two of Nim's three full-time teachers were male graduate students, Dick Sanders and Bill Tynan. During the previous year, some of Nim's female teachers had begun to wonder if they were physically strong enough to deal with Nim when he acted up and had to be punished.

In an all-out physical confrontation, there was no question that a male teacher had a better chance of subduing Nim than did a female, and it seemed likely that a man's relatively larger stature might somewhat inhibit Nim's aggressive tendencies. I felt strongly, however, that to worry about the outcome of a physical confrontation with Nim was negative thinking. A good teacher, male or female, should be able to avoid such a confrontation by anticipating trouble and by surprising or outsmarting Nim if he began to act aggressively. Anyone who had spent a fair amount of time with Nim should have been able to read his moods, particularly when he was not cooperating, and head off an outburst by changing activities. But if a teacher, male or female, failed to respond to his warning signals and allowed a physical confrontation to develop, physical strength and stamina often proved helpful in subduing Nim.

That physical size *per se* was not the crucial variable in working with Nim was apparent from the success of Laura, Susan, Joyce, and other female teachers, none of whom was very large. Instead of trying to have Nim obey her through the application of brute force, Laura dominated Nim by generally being one or two steps ahead of him. Like so many of Nim's other teachers, Laura also learned that showing Nim that she didn't love him or that she was angry at him was a much stronger weapon. Once a teacher established psychological and emotional dominance over Nim, a small measure of physical force could go a long way. It was not uncommon to see Nim stop in his tracks as he was about to misbehave after Laura and other female teachers raised their hands or simply signed *angry* or *you bad*.

All but two of the teachers listed in Appendix B had normal hearing. The major obstacle to attracting more deaf people to the project was a lack of funds. All of the inquiries I had from deaf people were channeled through deafness centers in the metropolitan area. None of these people could afford to work as volunteers. It was also difficult to find deaf people to attend to the esoteric question "Can a chimpanzee learn sign language?" when they felt that there was a crying need for their services as interpreters, counselors, and researchers on problems of deaf people.

Only four deaf people tried to qualify as Nim's teachers. Pat Dobro had worked as a deaf volunteer from January through April 1975, when Carol Stewart was in charge of the Columbia classroom. After a number of sporadic visits it became apparent that Pat had difficulty following Carol's method. As a result she dropped out of the project. The remaining three deaf people applied only after the project's future was stabilized by a grant from the National Institute of Mental Health in September 1976. Each of these applicants was interested in full-time work at a scale considerably higher than I could afford, but I was able to persuade them to consider part-time work and a free room in Delafield. One applicant gave up because he was unable to confront Nim and establish dominance over him. Another eventually decided that even with free rent he could not earn enough. The third applicant proved to be one of Nim's best teachers.

Mary Wambach, who did not lose her hearing until she was thirteen, provided the clearest evidence of how positively Nim responded to the fluency of a skilled signer. When I watched Mary sign with Nim, I saw a chimpanzee enthralled by the fluid movements of her hands. I had previously seen Nim show the same kind of fascination in signing when he was visited by highly skilled interpreters of sign language, hearing persons whose fluency in sign language was indistinguishable from that of a deaf person.

Mary's skill as a teacher did not derive solely from her fluency in sign language. She was also a skilled actress and a very unsentimental reader of Nim's moods and behavior. In many respects Mary's technique reminded me of Laura's. I could not help feeling that had Mary joined the project when Laura did (when Nim was only five months old), his signing would have progressed much more quickly. Mary's presence would have added to both the level of signing and the consistency of the approach she and Laura followed intuitively. Unfortunately for Nim, Mary did not meet him until he was three and a half years old and a veteran of many classes and many different teachers. Instead of meeting a helpless infant, Mary was confronted with a fairly independent chimpanzee approaching the period of his life that primatologists refer to as the "juvenile" (delinquent?) stage.

With a minimum of training Mary began to engage Nim in lengthy conversations. As I watched their conversations it was clear that Mary had none of the hesitation about her signs that was evident in many of Nim's nonfluent teachers. A good nonfluent teacher could manage one, two, or three sentences on a given topic before going on to sign about the next topic, but Mary could improvise at will and produce endless variations about any topic. After watching Mary work with Nim I was not surprised to find many interesting conversations in her notes. Excerpts from two of these conversations appear below.

Nim:	(looking at a magazine) *Toothbrush there, me toothbrush.*
Mary:	*Later brush teeth.*
Nim:	*Sleep toothbrush.*
Mary:	*Later . . . now sit relax.*
Nim:	(seeing a picture of a tomato) *There eat. Red me eat.*
Mary:	*There more eat! What that?*
Nim:	*Berry, give me, eat berry.*
Mary:	*Good eat. You have berry in house.*
Nim:	*Come. . . . There.*
Mary:	*What there?* (Nim leads me into house.)
Nim:	*Give eat there, Mary, me eat* (at refrigerator).
Mary:	*What eat?*
Nim:	*Give me berry.*
Mary:	*See rain outside?*
Nim:	*Afraid. Hug.*
Mary:	*You afraid noise?*
Nim:	*Mary, afraid. Hug.*
Mary:	*What you think about now?*
Nim:	*Play.*
Mary:	*What play?*
Nim:	*Pull, jump.*
Mary:	(later) *You tired now?*
Nim:	*Tired. Sleep, brush teeth. Hug.*

Male or female, hearing or deaf, Nim's teachers constituted a special group of people. In their own ways they were strongly dedicated to teaching Nim to sign and to insuring that his signs were well formed and appropriate to particular contexts. Of equal importance was their readiness to deal with all sorts of emergencies, whether it was filling in for a teacher who had to miss a session because of a last-minute problem or helping me prepare progress reports or buying and making new materials for Nim's classroom.

Nim's teachers often had to sacrifice large amounts of personal time in working on the project. They each had their own reasons for extending themselves, in most cases without any pay, but there was one common feeling that I sensed in all of Nim's teachers. Each seemed to derive a strong sense of satisfaction from getting to know and communicate with another species. That experience inspired many of them to strive toward professional standards of diligence and reliability that in many cases had never been demanded of them before. Working with Nim served as a way of fulfilling their own potentials.

9

Project Director
or Project Father?

From the very start of the project friends kidded me about being Nim's "daddy." After all, I had no children of my own. But whatever pleasure I derived from spending time with so appealing a creature as Nim, I always regarded him primarily as the subject of an experimental study. Of course I felt more affection for Nim than for the pigeons and rats with whom I had worked for the previous seventeen years, but seldom did those special feelings overshadow the true reason for my relationship with Nim.

Even if I had chosen to regard Nim as a surrogate son, a number of unexpected demands on my time would have made that impossible. I had a much harder time raising money for the project than I anticipated. Also, some of the volunteers required more time and attention than I had allowed for. Especially draining was the pressure I felt to reassure volunteers that they were working well. And there was the problem of interaction between certain volunteers, each of whom was trying to show me how special he or she was. One volunteer would put down the work of another, or, worse still, accuse other volunteers of maltreating Nim. If indeed there was a sense in which I was to be regarded as Nim's father, it would really be as *paterfamilias* of an often unruly family, bread-winner, listener, comforter, and peacemaker.

When I began Project Nim I was quite aware that it would pose two challenges: to organize and finance the care and education of an infant chimpanzee, and to master a number of new areas of psychology. But I had no reason to believe that the demands of this project would be so different from those I had experienced during fifteen years of organizing and raising money for research in a number of different areas of animal learning. If anything, I thought Project Nim would be a welcome relief. I expected that much of the time I would spend with Nim during his first few years would be devoted to play. I felt too that my dealings with Stephanie and her family and with other teachers would not be devoted entirely to the theoretical and scientific questions that motivated the project. I regarded both of these diversions as positive. Why not enjoy playing with the subject of your experiment, an activity that is not very rewarding when the subject is a pigeon or a rat? And why not temper intellectual discussions about Nim's use of sign language with anecdotes about his latest antics?

Even discounting Nim's unique personality and the subject matter of the project, it still had very little in common with other projects I had directed. In my earlier research I was always assured of the money I needed for personnel and equipment before I started a project. And there was never a need to work with pigeons and rats eighteen hours a day, 365 days a year. Nor were any special full-time caretakers needed. I could always leave my laboratory comfortably, knowing that my pigeons and rats were secure in their cages. In most cases, the experiments I performed could be interrupted with little or no consequence. If an interruption did cause a problem, I simply started again with new subjects. When an experiment did not succeed, it was an easy matter to repeat it with whatever variation was required.

All of these options were unavailable to Project Nim. Every unseized opportunity to teach Nim to sign seemed to be an opportunity lost forever. Even more than a human infant, Nim needed constant contact and attention. The fact that it was difficult to obtain funding could not serve as an excuse to postpone or to stop the project until funding could be obtained. Nor could the frustration encountered in finding the kind of dedicated and specially qualified workers needed, or problems in locating suitable living and working space.

When I decided to adopt Nim I knew there was no turning back. I was committing myself to an indefinite period of continuous research. What I didn't know was that I was also committing myself to a four-year search for funds and teachers while simultaneously trying to perform the research. For more than ten years I had been fortunate enough to receive ample and uninterrupted support for my research in animal learning and to have all my grant applications funded the first time they were submitted. The only difficulty I experienced in obtaining support were occasional reductions in the level of funding I requested. I was aware, of course, that during the period in which I received support, there was considerable expansion of federal funding for research in experimental psychology, but even though there had been some cutbacks in the budgets of the agencies that generally funded my research, I assumed that the obvious importance of studying language in chimpanzees and my previous record as a researcher made it likely that I would obtain some sort of support.

Nim was barely a half year old when I got my first inkling of the difficulties in store. An application for a small grant to the National Institute of Mental Health was rejected. A small grant is a one-year grant limited to $6,000, and its only advantage is that it comes through much faster than larger grants. Six thousand dollars was certainly not enough to pay for salaries on the project, but it could pay for food, diapers, and some much needed videotaping equipment. I also thought that obtaining a small grant would be an important step toward obtaining a grant large enough to cover all our expenses over a two- to three-year period.

The project's financial needs dictated that we reapply for funds immediately. It could not survive if we postponed submitting a new application until we obtained significant data on Nim's use of sign language. I felt that the best way to improve our original proposal, which had been criticized (rather unfairly, I thought) for a lack of detail about earlier work by others and about what we intended to accomplish, was to reveal clearly the deficiencies of current and completed studies on language in chimpanzees and to indicate how we would try to go beyond the findings of those studies. This strategy produced two related problems. First, it meant criticizing the work of precisely those researchers who would probably be asked to review the proposal. And second, though a central assumption was that socialization was an essential ingredient of our program for teaching language to a chimpanzee, there was no way to guarantee that our approach would result in Nim's using sign language as we hoped: spontaneously, motivated mainly by social reward, and as comments about important features of his world.

There was too much, in short, that we could not predict. Nevertheless, we proceeded to prepare an application for full support for a three-year period. This time our survey of the literature was given high marks, and our ambitious plans for socializing Nim and for documenting his use of sign language were applauded. But again the reviewers did not think it worthwhile to fund a massive effort to teach language to a single chimpanzee who had yet to utter the kind of sequences we wanted to study. The rejection came in December 1975. At that time Nim was two years old and had a vocabulary of forty-two signs. He had also begun to combine signs, but it would be some time before we could expect to obtain data that would differentiate our results from those of other projects.

For me the main consequence of having another proposal rejected was the ordeal of spending another three months of doing nothing but preparing a new proposal. As the focus of our next proposal, I planned to use Progress Report I, a summary of Nim's vocabulary and output of combinations through January 1976. I planned to include a number of photographs of Nim making various signs in which his only reward was the teacher's praise. In many cases Nim was shown signing about pictures of objects (see page 145). Also included was a sequence of photographs showing the stages Nim went through in acquiring a new sign (see page 141). By January 1976 I had collected a corpus of approximately 1,800 two-sign utterances, which enabled me to perform some simple analyses of Nim's grammatical competence. I noted, for example, that Nim signed *give me* and *give ball* more frequently than *me give* or *ball give* (see Chapter 11). For the first time there was a nonanecdotal basis for showing regularities in the combinations of a chimpanzee's utterances in sign language.

The reaction to this proposal was more positive than to our earlier proposals but funding was still not forthcoming. The National Science Foundation rejected the proposal outright, while the National Institute of Mental Health accepted it with a priority too low for funding. Had we not been kept afloat by grants from the W. T. Grant Foundation and the Harry Frank Guggenheim Foundation, it would have been impossible for the project to survive. There was barely enough money to pay for a full-time teacher, Nim's food, clothing, housing, a part-time secretary, and some videotapes that had to be used with borrowed equipment. From my own pocket I had to pay for vital part-time work that could not be obtained from volunteers—transportation, photography, xeroxing, educational toys, and other odds and ends. I certainly did not like the idea of digging deeper into my own savings in order to cover the costs of the project, but I was equally concerned about two other consequences of not getting funded: continuing the project with a staff of volunteers, and having to write yet another grant application and another progress report. As a result I would have to endure another long period during which there would be far too little time to spend with Nim.

I decided that I would write only one more proposal. The costs of running the project seemed to be growing geometrically as Nim became bigger, stronger, and more agile. His growth necessitated costly modifications of the Columbia classroom and his quarters at Delafield. And as Nim's use of sign language increased and he became stronger and more independent, the repeated introduction of new volunteers made it more difficult to tap his full linguistic potential.

There was little time to shake off the rejection of the latest grant applications. If I were to meet the next deadline, a new, up-to-date application would have to be prepared by early 1976, barely a month after the news that our previous applications had been rejected. Between Progress Report I and April 4, 1976, the closing date of data to be included in Progress Report II, there was a fourfold increase in the number of types of combinations Nim uttered. There was a similar increase in the sheer number of combinations of all types. For the first time we had a suffi-

ciently large body of utterances to demonstrate numerous regularities
in Nim's combinations. For example, Nim signed *more* + X (where X
stood for a wide variety of objects and actions, such as *hug, play, apple,
banana, eat*) rather than X + *more*; verb + *me*, as opposed to *me* +
verb, and so on. In Progress Report II, I also described Nim's simultaneous use of two signs and contractions of signs. This was the first
report of such combinations in a chimpanzee. Gratifying as Nim's linguistic progress was, its documentation required elaborate and time-consuming data analyses. Progress Report II, a single-spaced, fifty-seven-page
document, was completed with literally minutes to spare and sent off as
a supplement to our new applications to the National Science Foundation
and the National Institute of Mental Health. Decisions from both of these
agencies were expected by early June.

On June 18 I mustered enough courage to call the granting agencies.
Again NSF rejected our proposal. But shortly after this I received an unexpected call from Dr. George Renaud, executive secretary of the study
section that reviewed my application to NIMH. Instead of saying
definitely yes or definitely no, Dr. Renaud said maybe. Apparently enough
interest had been generated by the latest application to warrant a closer
inspection of Project Nim. For this purpose Dr. Renaud wanted to
arrange a "site visit" by a group of experts who would observe first-hand
how and what Nim was signing. The report of the site visit committee
would determine whether the project would be funded.

I was quite prepared for a definite acceptance or rejection from
NIMH. Their maybe threw me completely off balance. By the end of
June 1976, I had expected either to push ahead and try to realize the
goals of the summer or, if both applications were denied, to end the
project immediately. The site visit, scheduled for the first week of August,
simply intensified problems that had long since become almost unbear-
able. Extra funds would have to be found to continue to pay minimum
expenses for at least another month. It would also be necessary to prepare
a Progress Report III to bring the committee up to date on Nim's use of
sign language between April 4 and the beginning of July.

There was no choice but to continue. Within a day of learning about
the site visit I began to organize the effort needed to put on an impressive show. Just about everybody on the project felt that we would
certainly get funded if NIMH was willing to send experts to observe
what Nim could do. I was less optimistic. I knew of enough instances in
which site visit committees had rendered negative verdicts.

The challenge of preparing for the site visit unified members of the
project as never before. I could not help wondering whether the kind of
work that the volunteers performed in preparing for the site visit would
ever have been done by salaried employees. Once I outlined my plan

for the site visit during the last week of June, no one took a day off until August 3, the day of the visit.

Before bringing the committee to the observation room, I planned to show them some videotapes of Nim signing with different teachers. These videotapes would be shown in slow motion and would be accompanied by a transcript of what was being signed. This would not only familiarize the committee with the nature of the communication they would see between Nim and his teacher, but it would also give them a first-hand glimpse of the detailed records we kept of Nim's signing. Laura and Andrea Liebert worked around the clock transcribing some of their recent videotapes. They then transcribed each other's tapes in order to check the reliability of their original transcripts. A further measure of reliability was provided by the transcripts of an interpreter of sign language who had never seen Nim sign. When she was done, it turned out that her readings agreed with more than 90 percent of the entries in Laura's and Andrea's transcripts.

I was unprepared for the tension a site visit can produce. A few days before the visit Dr. Renaud outlined the program. First there would be a general discussion of the issues presented in the grant application. In effect, this was an informal oral examination for myself and Tom Bever, the co-principal investigator. This would be followed by demonstrations of Nim's signing, first via videotapes and then by observation of a classroom session. The committee would then meet by themselves and determine if they needed any additional information.

The first part of the visit went smoothly. Tom and I both felt we had fielded the questions put to us without much difficulty. The tapes and the transcripts of the tapes were well received. Laura had as good a session with Nim as I had ever seen. Dr. Harry Boernstein, a committee member who was a fluent signer, interpreted what Nim was signing for the others. In the space of a half hour Dr. Boernstein observed Nim make more than forty signs and at least that many different combinations. Nim was very attentive to Laura throughout the session despite the fact that he could hear the whisperings and rumblings of the committee in the observation room. Nim was also clearly uncomfortable under the floodlights I had turned on to make the classroom easier to see through the dark one-way window. It was encouraging to see various members of the committee congratulate Laura on Nim's performance when they met her later during a lunch break.

A few weeks after the site visit, Dr. Renaud called to tell me that a two-year grant of approximately $140,000, plus overhead, had been approved At long last Project Nim was given the recognition and support it needed to survive.

Too soon after hearing the good but unofficial news from NIMH, I began to see a number of ominous clouds. In June, I had promised to give Laura, Amy, and Walter a definite decision about jobs and salaries for the coming academic year. They had each warned me that they might have to leave the project in order to advance their careers or find jobs offering more security. Each time I talked them out of leaving by hoping out loud that a full-time job and an opportunity to experience the fruits of their labors with Nim would provide enough of an incentive to continue. I realized that I could not keep this up indefinitely. By June firm word from me about what I could offer each of them was long overdue. This was especially true in the cases of Laura and Walter, who felt strong pressure to continue their educations in September. The delay imposed by the site visit forced me to postpone my decision about the future of the project. Unfortunately, Laura, Amy, and Walter had passed the point beyond which they could delay their decisions.

Laura, who had already postponed graduate school for a year, had recently been accepted into the linguistics program of the University of California at San Diego. She was also offered a research assistantship at the Salk Institute by Dr. Ursula Bellugi, an internationally recognized expert on the development of sign language in children. Amy Schachter had an offer of a full-time job working with primates at the Bronx Zoo. Walter Benesch had been accepted by the School of Social Work of Boston University and could no longer postpone a request to delay the beginning of that program for a year in order to continue working on the project. For a number of months I had been aware that the fourth original occupant of Delafield, Andrea Liebert, would leave at the end of August, to get married and then enter graduate school at the University of Tennessee.

My elation over the news that the project would be funded sometime between September and December quickly gave way to despair. Within days of learning that the project was financially solvent I was faced with a managerial problem unlike any I had experienced previously: the task of replacing so many key people all at once. Even when Nim had moved to Delafield, there had been ample time to arrange for him to get to know his new caretakers before the move. And then, Nim had been a year younger and more amenable to forming bonds with new people.

The group I now selected to work with Nim was highly talented and dedicated. Bill Tynan had been on hand during the previous two years, ever since Nim had begun his classes at Columbia. Bill was scheduled to begin work as a graduate student in Columbia University's department of psychology in September, and planned to use his work with Nim as a basis for an M.A. thesis. He agreed to become one of Nim's resident caretakers at Delafield. Bill was a model of the kind of teacher I had been seeking for the project. Joyce Butler, a recent college graduate who had performed well as a volunteer during the previous nine months, agreed to continue on the project in two roles. As a resident caretaker at Delafield, she assumed the vital role that Laura had initiated: coordinating activities at Delafield with those of the classroom. Joyce also agreed to continue as a regular classroom teacher. Susan Quinby, another of Nim's accomplished teachers, agreed to continue taking sessions in the Columbia classroom, but had to cut back somewhat because of her full-time job teaching children with learning disabilities. Dick Sanders, a newcomer with impressive credentials, asked to work on the project a day after Laura made a firm decision to start graduate school in California. Prior to his graduate training at Columbia, Dick had assisted David Premack in his study of the ability of a chimpanzee to learn an artificial language. Dick was the first project member to have a solid background in psycholinguistics. All that remained for him to do in order to earn his Ph.D. at Columbia was to turn in a doctoral dissertation, a dissertation that would analyze Nim's use of sign language. Even though Dick didn't know sign language when he joined the project, he quickly learned. His reliable and careful work in the classroom and his intellectual leadership proved to be invaluable.

To the extent that Dick, Joyce, Bill, and Susan were available, I was satisfied that Nim was receiving sound instruction in sign language. Yet as talented as this group was, they could hardly be expected to solve a major problem: Nim's depression over the abrupt loss of his previous teachers. Also, since Dick and Joyce were the only full-time people on the project, they were under constant pressure to pick up extra sessions, pressure they understandably resisted and resented. Dick wanted to start his dissertation as soon as possible and needed to spend more time with his family. Joyce was assuming too many responsibilities, and often ended up working seven days a week.

The first clear signal of impending financial trouble came from NIMH. After a second site visit I was informed that even though the committee was satisfied with Nim's progress, constraints on the NIMH budget made it impossible to award an emergency supplement. A similar appeal for a supplementary grant from the Harry Frank Guggenheim Foundation also failed. These rejections left us, at best, with enough money for a small staff of teachers and some baby-sitters. I still could not afford to hire a full staff of adequate teachers.

Within a few months of receiving the grant that I had been struggling to obtain for three years and that I had expected would provide the project with financial security, I had to come to grips with the fact that it was simply too late. That grant presupposed a continuity of Nim's major teachers. After Laura, Amy, and Walter left, it became clear that the long-awaited funds could never compensate for the loss of such a stable and talented "family." Watching Nim's failure to adjust to this loss, I realized that the days of the project were numbered unless I could quickly come up with enough money for additional teachers who could work with a chimpanzee who was often too distracted to cooperate.

Giving up on foundations, I tried to interest a professional film maker in making a television documentary about Nim. This almost succeeded— one prominent film maker, Robert Drew, an independent television producer, very nearly managed to raise the requisite funds but in the end the networks could not be convinced. The same thing happened with several other film makers. Finally, after many false starts, I reached an agreement with the Children's Television Workshop, which wanted to film Nim as a character for its "Sesame Street" program. They regarded Nim as an excellent role model for showing young children how to eat, get dressed, use the toilet, look at picture books, draw, and so on. The Children's Television Workshop agreed to underwrite a substantial portion of the cost of the project while the filming took place. This support provided momentary relief, but it did not allow me to achieve the goal that was rapidly becoming essential: to stop using volunteer teachers and arrange for Nim to be taught exclusively by a permanent staff of professional or preprofessional teachers.

Of all the problems arising from our failure to get a grant until Nim was almost three years old, the most serious was my dependence upon volunteers to look after and to teach him. As talented as those volunteers were, they were the source of many difficulties. When a volunteer left the project, a large investment in time was needed to recruit and train a replacement. For me and the other teachers, that time could have been spent more advantageously in teaching Nim and in analyzing what he was signing about.

Less obvious, but worth noting for the sake of anyone contemplating a similar project, were two only slightly less vexing problems: the nonprofessional attitude of many volunteers, and the sheer number of people I had to cope with on a day-to-day basis. Instead of talking about the intellectual and scientific questions suggested by their work, too many volunteers preferred to talk—to me—about various sorts of personal problems. I could have evaded such discussions by claiming that I was too busy. But I could not help recognizing that I owed something to the volunteers who worked for me. Aside from a friendly ear, there was little else I could offer. It also became quite clear that if I was unresponsive to appeals for attention from volunteers, their involvement in the project would inevitably decline.

It was especially upsetting to have to put aside compelling work on a grant proposal or on a progress report in order to settle arguments among members of the project. I was painfully aware of the fact that when a group of from six to twelve teachers all worked with one chimpanzee, it was almost unavoidable that there would be competition over who was the most effective teacher, who loved Nim the most, and whom he liked the best. In some instances disputes grew out of complaints that particular teachers were lax in accepting certain kinds of behaviors and ways of signing from Nim. There were complaints that some teachers taught too few sessions, or that they did not clean up after a session in the classroom, left a mess at the house, or didn't report something that was broken in time to have it repaired for another teacher to use. Most of the complaints were valid, but I couldn't help thinking that they were sometimes motivated, at least in part, by a desire on the part of the complainer to be one up on the others.

These problems forced me to assume a role that I never bargained for: acting like a father to a frequently squabbling group. Too many volunteers failed to grasp that the endless one- to two-hour interruptions spent in talking to upset, annoyed, or overcompetitive teachers left me with too little time to do anything else. Try as I might, I was unable to instill in enough of them a professional attitude toward the project.

At the beginning I had responded to various unprofessional demands with good humor, assuming that I would only have to work with volunteers for the "time being." The "time being" was always defined as the interval between writing a grant proposal and getting it funded. But with each rejection, I became more frustrated by the need to run the project on a basis that did not satisfy my professional standards.

Under the circumstances, it was difficult to respond to legitimate demands from such outstanding teachers as Laura and Bill. They were not only able to work autonomously, but they often went out of their way to patch up disputes so that they would not require my attention. Under normal circumstances I would hasten to recognize such efforts. On Project Nim, however, where the extra efforts of many people were needed and there was a steady procession of deadlines for grant applications and training new volunteers, there simply was not enough time to recognize good work or to discuss the project with those teachers with whom I could share a professional point of view.

There was a certain irony in the fact that on the rare occasions when I was not too emotionally drained to be able to talk to close friends whose support and interest I would normally have welcomed, it often proved difficult to get them to understand why the project was so frustrating. To outsiders, the most salient feature of Project Nim was the fascinating possibility of communicating with a lovable member of another species. Surely, they felt, it was worth enduring a few difficulties to achieve that kind of communication. And besides, wasn't it fun to teach my "son" how to sign?

The answer, of course, was yes. I would have liked nothing better than to assume the role of Nim's father by being his teacher and care-taker, especially had I been able in the process to avoid the other responsibilities of fatherhood imposed on me by the nonscientific demands of the project. One of the many things I learned from Project Nim is that if one's aim is to be father to a chimpanzee, one should be a rich father with as small, stable, and professional an extended family as possible.

10

Nim's Vocabulary

Whatever Nim learned about sign language, he learned because of the hothouse environment of signing he experienced both at home and in his classroom at Columbia. Just how Nim derived his knowledge of sign language from this environment is a question that is difficult to answer for two reasons. Psychologists and linguists have yet to agree on a theory of how children learn to speak. Volumes of data notwithstanding, definitive answers are still lacking to such basic questions as what kinds of grammatical rules best describe a child's initial utterances and to what extent a child's grammar is learned or is inborn—the manifestation of an innate "language acquisition device." But even if a definitive theory of child language were available, it would be rash to assume that a single chimpanzee, no matter what its use of sign language, had learned to use language as a child does. At present, there is no justification for going beyond a description of how we went about teaching signs to Nim and what Nim learned about sign language.

These facts, however, are exciting. Nim's knowledge of sign language is best described on two levels: his vocabulary of single signs and his tendency to combine these signs into sequences of signs. In this chapter I will describe Nim's vocabulary of signs, how he learned them, and the humanlike ways in which he used them. Having considered the range of topics Nim signed about with single signs, we will be in a better position, in the next chapter, to understand the sequences of signs he emitted and to ask whether these sequences were sentences.

Nim learned to express 125 signs during his first forty-four months. A description of the usages of these signs and how they are made can be found in Appendix C. Nim mastered his first sign, *drink*, at the young age of four months and went on to acquire five other signs (*sweet*, *up*, *give*, *more*, and *eat*) during the next four months. As impressive as his progress may seem, it is important to keep in mind that the circumstances of Nim's education in sign language were far from ideal. During the period shown, Nim was taught by more than sixty teachers. Two to three times as many prospective teachers also spent at least one session with Nim trying to demonstrate that they could qualify as regular teachers.

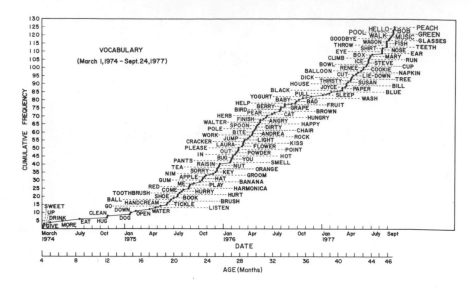

The sequence in which Nim learned new signs: the slope of the graph shows his rate of learning.

Emotional disruptions in Nim's life caused by the necessity of replacing volunteer teachers were not the only factor that contributed to the irregularities of the learning curve shown in the figure. After Nim's introduction to his classroom at Columbia in September 1974, he needed a few months to adjust to his new environment and to the activities introduced by Carol Stewart and other teachers. The classroom was his first extended experience outside a home environment. For these reasons, it is not surprising that his initial rate of progress in the classroom was not rapid.

In June 1975, Laura Petitto took over the supervision of Nim's instruction in the classroom. At that time, Nim was one and a half years old. The combination of Laura's talent, Nim's adjustment to his classroom, and his accelerating intellectual development produced a dramatic increase in his progress. Between June 1975 and May 1976, Nim's rate of learning new signs more than doubled: from two to five signs a month. During this time spirit on the project peaked. Subsequent fluctuations in the rate at which Nim learned new signs resulted from the need to give his regular teachers well-earned vacations and from his slow adjustment to new teachers. It is not surprising that the slope of Nim's sign-learning curve decreased sharply in September 1976 (following the departure of Laura, Walter, Andrea, and Amy). It did not reach its former level until the following summer. During September 1977, Nim's last month in New York, he was learning new signs at a rate of two per week. While Nim was adjusting to new teachers, his general level of signing decreased markedly. Too much time was spent engaging in more basic forms of communication with what must have struck Nim as an endless stream of candidates seeking to fill vacant teaching positions. Instead of signing with new teachers, Nim acted like a brat and tried to avoid familiar and previously enjoyable tasks in the classroom.

In the absence of norms describing the linguistic development of a chimpanzee, it is impossible to judge just how Nim's use of language would have grown had his original caretakers remained at Delafield. It seems reasonable to assume, however, that there would have been a steady increase in the rate at which he acquired new signs. It also seems clear that much of Nim's antisocial behavior, which obviously interfered with our ability to sign with him, could have been avoided. Before Laura, Amy, Andrea, and Walter left, Nim was docile and relatively easy to control. A year later, there was no one on the project who could do more than hold him at bay when he began to be uncooperative. Undoubtedly the loss of Nim's immediate family at Delafield at a critical stage of his growth had a permanent adverse effect on his social, linguistic, and emotional development.

Nim learned most of his signs by one of two simple methods. Either his teacher molded his hands into the correct configuration, or the teacher simply made the sign for Nim to imitate. Less tangible than the methods of molding and imitation was the motivation we tried to provide for Nim to sign. By signing all the time, Nim's teachers tried to get him to understand that signing was a way of communicating and that he would be left out of this kind of activity if he didn't sign. Our goal was to make Nim want to sign in order to be included in what his teachers were doing.

During the first few months in his Columbia classroom, Nim was not terribly responsive to his teachers' efforts at molding his hands to make different signs. However, as Nim added more and more signs to his vocabulary, he began to give his hands to his teacher to mold, particularly when it became clear to him that he did not know the sign for some object or event in which he was interested. Eventually, only a few molding demonstrations were needed before Nim learned to make the sign on his own. We interpreted Nim's offer of his hands to be molded as a primitive way of asking *what?* and as a positive outcome of our efforts to motivate Nim to sign.

In many instances it was possible to teach a new sign through imitation alone. Laura, for example, taught Nim to imitate her *orange* sign after looking at a picture of an orange. Dick taught Nim to sign *tree* by pointing to a tree and signing *tree*. Nim showed his understanding of the sign by signing *tree* himself and then running around the tree or jumping up into one of its branches.

A representative example of how Nim learned a sign at an early stage of the project is provided by the details of Laura's notes on the acquisition of *tea*. In this and other instances, clear stages could be discerned in Nim's acquisition of a new sign. These can be seen in photo-

Learning signs through imitation: *orange* and *tree*

graphs I took of Laura and Nim. As he became older, many of these stages were passed through very quickly or omitted.

Stage One: Nim's attention is directed toward an object. In this instance, Laura poured some hot water into a cup containing a tea bag and then engaged Nim in the following conversation:

Teacher:	*You drink this?*
Nim:	*Me drink.*
Teacher:	*Drink what?*
Nim:	*Drink.*

At this point, Laura molded Nim's hands into the *tea* sign and pointed to the cup of tea encouraging him to take a sip.

Stage Two: Nim offers his hands to Laura. This is a clear request that Laura hold his hands so as to make the sign that describes the object in which he is interested.

Stage Three: Nim attempts to mold his teacher's hands to make the appropriate sign. The teacher of course cooperates. Once the teacher's hands have been molded by Nim, he is allowed to obtain the object he desires, in this case a sip of tea.

Top left: Laura molds Nim's hands into the sign *tea*. Top right: Nim approximates the sign; he then tries different signs he knows are relevant to the situation: *give* (center left), *drink* (center right), *more drink* (bottom left), and *more* (bottom right).

Stage Four: Nim attempts to approximate or imitate the sign with his own hands. Occasionally Nim would revert to other signs that he had learned were relevant to the situation.

Stage Five: Nim makes the sign correctly without molding or any physical prompting other than the teacher's making the sign in conversation, for example:

Teacher: *Laura drink tea* (teacher drinks).
Nim: *Me drink tea.*

Stage Six: Nim makes the sign without any cues from the teacher. For example, upon seeing the tea cup, Nim will sign, *tea, tea.*

Stage Seven: Nim's frequency of making the new sign increases sharply during the next few weeks. This stage has two important aspects. First, the sign occurs in both appropriate and inappropriate contexts. During a chase game, Nim might sign *Play me tea.* This development seems comparable to the observations of psycholinguists that a child "overgeneralizes" new words. For example, after a child learns to say

The next day, he signs *tea* spontaneously.

"doggie," it often goes around saying "doggie" to a wide range of objects, both animate and inanimate. Nim also "practiced" new signs: he was often seen gazing at his fingertips while making the sign in the absence of the sign's referent.

Stage Eight: At this point, the sign occurs mainly in the appropriate context. Nim signs *tea* only when tea is present or when he is specifically asking for tea.

With the exception of stage seven, which usually lasted at least a week, the duration of the other stages was quite variable. In some instances, Nim went through the first six stages in as little as thirty minutes (as he did in the sequence shown below), or he could take more than a month to learn to make a sign spontaneously.

Below left: To obtain the red balloon Bill is holding, Nim imitates sign *balloon*. Below right: This time, when Bill shows him the balloon, Nim signs *red*. Bottom: Later, he signs *balloon* in response to Bill's questions *What's that?* (left) and *Name?* (right).

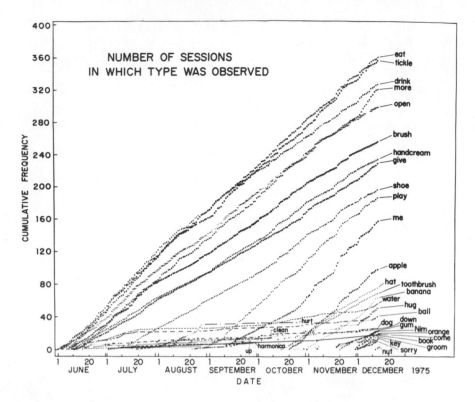

This graph shows the number of sessions in which particular signs were observed during an early phase of the project.

Since most of Nim's teachers were right-handed and he learned to sign by molding and imitation, it is not surprising that he made most of his one-handed signs with his right hand. When Nim didn't sign with his right hand, it was usually because he was using it for some other purpose, such as holding on to the teacher or some object with which he was playing. Under these circumstances, Nim signed with his left hand. He also signed with his feet! I was once tickling Nim when he was holding on to the branch of a tree with both hands. Nim indicated that he wanted to be tickled again by signing both *more* and *tickle* with his feet. On other occasions when Nim's hands were otherwise occupied, I have seen Nim point to an object he wanted with his feet in order to sign *that*.

Nim's day-to-day usage of signs was determined by his needs, the demands of his teachers, and the situations to which he was exposed. It was relatively easy to make sure that Nim signed a particular sign each day by arranging circumstances appropriate for that sign. Generally, once a sign was acquired, it occurred at least once during each session. The few exceptions (for example, *clean, hurt, ball, harmonica, up*) can be attributed to the absence of a demand that a particular sign be used.

For example, *harmonica* was prevalent during the summer that Bob Sapolsky, a volunteer teacher, worked with Nim. Following Bob's lead, other teachers also interested Nim in harmonicas (which he only later learned to blow into successfully). As a result, Nim signed *harmonica* on a regular basis. After Bob left, other teachers replaced the harmonica with other activities, and the frequency of *harmonica* decreased. *Hurt* was used only when Nim hurt himself or when he noticed a scratch or scar on someone else. As Nim became more active, he signed *up* and *down* less frequently. At this stage of the project, the signs *ball* and *clean* were called for only sporadically. Even though many of Nim's signs fell into disuse, they were easily restored. All that needed to be done was to provide the appropriate context and, in some instances, a few reminders as to how the sign should be made.

In deciding which signs Nim should be taught, I was guided by two considerations: what objects, actions, relationships, and attributes Nim was interested in; and what signs could be used readily in combination with one another. After Nim had acquired a substantial vocabulary of nouns, I emphasized verbs and prepositions, signs that I hoped he would combine with nouns to create interesting combinations.

Nim's main reward for learning to sign was our approval and being able to sign about something that was important to him. This was most evident in the ease with which we taught Nim to sign in response to pictures of objects and his tendency to thumb through picture books and magazines and sign about what he saw. Nim regularly identified pictures of food and drink even when he wasn't hungry or thirsty. In many instances, he declined a serving of the actual object he had just signed about when it was offered to him. Nim also signed about pictures of many non -food or -drink objects.

Nim often signed to pictures: *toothbrush* and *key*.

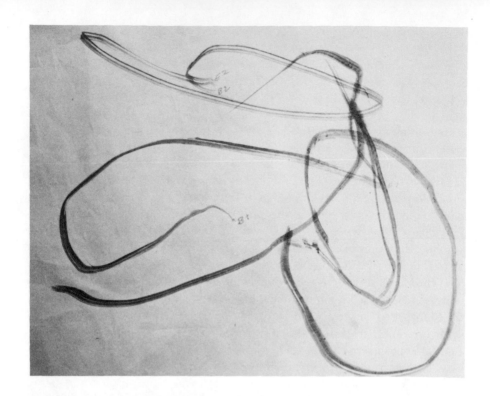

Above: a free-hand drawing by Nim. B1 and B2 mark the beginning and end of the first color used (red), E1 and E2 of the second (green).

Below: When shown a piece of paper with two circles on it, Nim would habitually join the circles up. Here, B marks his starting point.

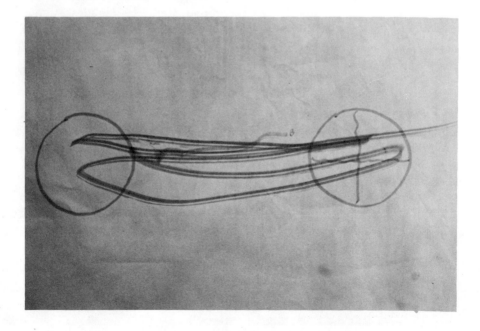

Nim at work

Thanks to the patient efforts of Bill Tynan, Nim developed a strong interest in drawing with a variety of colored crayons and paints during his first year at Columbia. Given a choice of crayons, Nim often showed preferences for particular colors, providing a natural situation in which to teach him their names. In order to be given a particular color, Nim had to sign its name. Often he was given a choice of two to five colors. If he signed incorrectly he was given either a nonpreferred color or nothing.

With the help of photographs, Nim learned his own name sign as well as those of his teachers. In sign language, name signs are made up by using the hand configuration that corresponds to the first letter of a person's name (see Appendix A). That hand configuration is used to identify some personal characteristic. For example, my name sign calls for the *h*-hand to touch my mustache. Laura's name sign calls for the *l*-hand to touch her forehead above her eyebrow as if she was engaging in her common habit of brushing her hair to the side of her face.

By the time we began to teach Nim his name sign, he was already signing *me* to pictures of himself. He also signed *me* to his image in mirrors. In the presence of pictures of himself, as well as when he was looking in the mirror, Nim's teachers molded his name sign, the *n*-hand scratching the side of his head. Once Nim learned how to make his name sign, his teachers regularly asked him *Who's that?* in the presence of his picture or image in the mirror. He replied by signing *Nim*.

Signing the names of colors: *orange* (left), *black* (right)

Signing the names of his teachers. Above left: *Laura*. Above right: Joyce: *Who?* / Nim: *Joyce*. Below left: *Bill*. Below right: *Dick*

Naming himself. Susan: *Who?* / Nim: *Nim*

Teaching Nim the name signs of his teachers was somewhat easier. One technique was to have one teacher point to another and sign the second teacher's name. The first teacher would then mold Nim's hand to form the name sign of the second teacher and also require Nim to make that sign before the second teacher was allowed to approach. Another technique was to have Nim's current classroom teacher hold up a photograph of the next teacher and sign *Who's that?* Since Nim was highly accustomed to transfers from one teacher to another and since he could often hear the new teacher from outside his classroom, it was easy for him to associate the photograph of the new teacher with that teacher. Before the new teacher would come into the classroom, Nim was required to sign that teacher's name.

Once Nim learned the name signs of various teachers, he often signed their names upon first seeing them. Dramatic examples of Nim's signing the names of his teachers were provided by reunions with teachers he hadn't seen in a long time. For example, when Laura came back to visit Nim after a twelve-month absence, one of Nim's first signs was *Laura.* When Nim greeted Walter, whom he hadn't seen in more than nine months, he signed *Walter.*

Certain usages of Nim's signs were quite unexpected. At least two of Nim's signs (*bite* and *angry*) appeared to function as substitutes for the physical expression of those actions and emotions. Nim learned the signs *bite* and *angry* from a picture book showing Zero Mostel biting a hand and exhibiting an angry face. During September 1976, Amy began what she thought would be a normal transfer to Laura. For some reason, Nim didn't want to leave Amy and tried to drive Laura away. When Laura persisted in trying to pick him up, Nim acted as if he was about to bite. His mouth was pulled back over his bare teeth, and he approached Laura with his hair raised. Instead of biting, however, he repeatedly made the *bite* sign near her face with a fierce expression on his face. After making this sign, he appeared to relax and showed no further interest in attacking Laura. A few minutes later he transferred to Laura without any sign of aggression. On other occasions, Nim was observed to sign both *bite* and *angry* as a warning. Nim also signed *sorry* after misbehaving, for example, after jumping around too much in the class-room or breaking a toy. On many occasions he signed *sorry* even before his teacher reacted to his transgression.

Nim's signing of *bite* and *angry* to express his feelings is the only example I know in which a chimpanzee or any nonhuman species has substituted an arbitrary word for a physical action. Both Washoe and Nim learned to sign *dirty* when they wanted to use the toilet and *sorry* when they wanted to placate a teacher who was upset about an episode of bad behavior. (It is of course possible that Washoe and Nim used *sorry* simply to avoid punishment and not because they were expressing true remorse. But you could say this about people too. The sign *dirty*, which should actually be glossed as *I have to go*, is a clear-cut example of a chimpanzee reporting a bodily state.) But reporting a state such as remorse or the need to go to the bathroom is quite different from signing about an impulse and, as a result of signing, no longer feeling the need to express that impulse. It is commonplace to observe that humans avoid actual physical aggression by warning someone of their feelings before they strike. Animals make threatening gestures, but these gestures are inflexible: as far as we know, they cannot be modified. Nim's learning to sign *bite* and *angry* as warnings and his tendency not to attack after perceiving that his warning had been recognized represents an important use of arbitrary signs by a chimpanzee. It suggests that they can learn to control their aggressive impulses by signing about them.

Another unexpected way in which Nim used signs was to misrepresent his bodily states. Once Nim was toilet trained he learned to sign *dirty* to indicate that he wanted to use the toilet. He also learned to sign *sleep* to indicate that he wanted to go to bed or to take a nap. Normally the sign *dirty* would be followed by a trip to the toilet. Having signed *sleep*, Nim would usually be allowed to take a nap or go to his bedroom.

Nim could not have failed to perceive that his teachers were particu-

Nim signs *dirty*. Joyce responds by signing *House?*—that is, "Do you want to go into the house?"

larly attentive to the signs *dirty* and *sleep*. In the case of *dirty*, his teachers were eager to avoid toilet accidents. Short of taking Nim to the toilet every half hour, the most reliable way to avoid a toilet accident was to encourage Nim to sign about his need to use the toilet and to respond to his signing immediately. Because Nim's teachers were concerned about his going to sleep on time and were aware of the significance of having a chimpanzee communicate about a bodily state, they were also very responsive to Nim's signing *sleep*.

It wasn't long before Nim began to sign *dirty* even when it was clear that he had no need to use the toilet. Within minutes of having urinated and/or defecated, he might sign *dirty*. Likewise Nim signed *sleep* while showing every sign of being fully alert. The misuse of *dirty* and *sleep* seemed to be motivated by a desire for change. If Nim was bored by what his teacher was having him do in the classroom, he might sign *dirty* or *sleep*. That he was bored could be perceived easily by his looking away from his teacher, running around the classroom, or otherwise resisting his teacher's efforts to focus his attention. Nim also signed *dirty* and *sleep* when he wanted to delay his transfer to a new teacher or to get out of working on some task.

It was often easy to tell whether Nim was using *dirty* or *sleep* in-

Nim is bored. He signs *dirty* to provide a diversion.

A "lying" *sleep* sign. Nim's eyes do not meet Dick's.

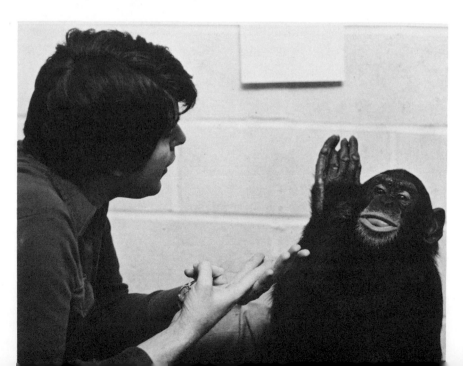

appropriately even in the absence of information as to when Nim last used the toilet or how tired he was, because he would usually avoid eye contact and often show a slight grin on his face. If the teacher responded to an inappropriate *dirty* sign by taking him to the toilet, Nim often used that opportunity to play by climbing the pipes or jumping from the sink. If Nim had misused *sleep*, it was obvious as soon as he lay down. He would coyly sneak looks at his teacher to see whether he was being watched. If the teacher had decided to use Nim's request for sleep as an opportunity to take a breather, Nim often jumped away and tried to capitalize on a lapse in the teacher's attentiveness by running around the classroom.

When Nim's teachers felt that he was misrepresenting his condition, they often indicated that they were not fooled. Typically they replied to Nim's false sign by signing *you not dirty* or *you not sleepy*. Then Nim would switch to another tactic in order to get free of his teacher. He might point to the door and sign *open* or *me out*.

When *dirty* or *sleep* were in fact appropriate, Nim's reactions to his teachers' challenges were quite different. If Nim really had to go to the bathroom and his teacher didn't respond to his initial *dirty* sign, Nim repeated the sign, often with both hands. On some occasions, Nim underscored the urgency of his request by pulling off his pants. When Nim really wanted to go to sleep after signing *sleep* and his teacher did not respond, he either signed *sleep* repeatedly or he signed other appropriate utterances, for example *hug* (to say that he wanted to be taken up to his bedroom) or *hurry*.

Nim's behavior when his teacher did not honor a valid *sleep* sign was revealing. If Nim asked to go to sleep too early, his caretaker would usually ignore the request because Nim would wake up too early the next day. His morning caretaker rarely came to get him up before 9:30, and if Nim got up much before that and had to wait impatiently for his caretaker to unlock his bedroom, he would act for the rest of the day as if he had gotten out of the wrong side of his bed. When Nim had given up trying to convince his caretaker to put him to bed by signing *sleep*, he often curled himself up and went to sleep in the lap of his caretaker. There was many a night when Nim had to be carried to his bedroom sound asleep.

In other studies of apes and sign language, the use of *dirty* in situations where its usual meaning does not apply has been interpreted differently. Roger Fouts has referred to some of Washoe's utterances containing *dirty* as examples of cursing. Washoe signed *dirty monkey* after a macaque threatened her and while she was being drilled about the names of the gibbons and the macaque. (The correct response was *monkey*.) She also signed *dirty Roger* after Roger Fouts refused to let her out of her cage at the Oklahoma Institute for Primate Studies. Francine Patterson reported that Koko (a gorilla) used the "expletive" *dirty* after being accused of damaging a doll that was first damaged by Michael, a younger subject of Patterson's study of signing gorillas. According to Patterson, Koko was "aware that she was only 50% guilty." That was why Koko ". . . retaliated with the worst insult in her lexicon": *you dirty bad toilet.* Neither Fouts nor Patterson explained just how their apes acquired the concept of cursing. I would like to suggest that in these and in other instances of presumed cursing Washoe and Koko were each trying to get out of unpleasant situations. What better strategy than to use a sign that results reliably in an immediate move to a new situation?

With a few exceptions, all of the signs in Nim's vocabulary were modeled after the signs of American Sign Language. In two instances, Nim invented his own signs (*hand cream* and *play*). In some instances (for example, *groom*), there was no standard sign in ASL to use as a reference. Other signs (for example, *blue* and *green*) were too difficult for Nim to learn. Accordingly, we modified these signs into simpler forms.

We had no plan to teach Nim to sign *hand cream*. When he invented that sign, we were trying to teach Nim the signs *brush* and *groom*. At the beginning of each day in the classroom, Nim was brushed and groomed by his teachers. Since Nim enjoyed these activities, it was not hard to teach him to sign *brush* and *groom*. During this portion of the session, Nim's teachers occasionally rubbed some cream on Nim's hands out of concern that they might be too dry. Unexpectedly, Nim showed much interest in hand cream, even when his hands seemed amply moist. Spontaneously, he began to rub his hands together, apparently in anticipation that his teacher might provide him with another dab of cream.

Before deciding if we should teach Nim a standard sign for *hand cream*, we consulted Ronnie Miller, the only member of the project who had learned sign language as a first language, along with other experts on sign language from the NYU Deafness Research Center. We learned that there was no standard sign for *hand cream*, so we accepted Nim's sign of rubbing his hands together.

The circumstances under which Nim invented the sign *play* were somewhat different. Shortly after Nim was moved to Delafield, a group of his teachers gathered for a weekend picnic. While sitting around a tree in which Nim was playing, they began to clap their hands rhythmically. Nim seemed intrigued by this activity. After watching his teachers clap in unison, he jumped down from the tree and with a big grin began clapping his own hands together. What happened next is a matter of interpretation. Three teachers noted independently that Nim acted as if he was not content to simply be a member of a group of people clapping their hands. Instead, he seemed to want to direct everyone's attention

Above: He really has to go. He signs *dirty* with both hands. Below: He pulls off his pants to convince a skeptical teacher of his seriousness.

groom   *brush*

toward him. Still clapping, he ran off and looked back to the group, challenging someone to chase him. Within seconds Laura got up, still clapping her hands, and chased Nim around the tree. As was true on other occasions when his teacher caught him, Nim fell to the ground, rolled over, and asked to be tickled. As previously, Laura complied. Later during the same day, while alone with Amy, Nim spontaneously clapped his hands, ran off, and beckoned to her to chase him. Nim went through the same routine on a number of other occasions that weekend with other teachers. By the time we had our next staff meeting, on the following Wednesday, there were other reports of Nim clapping his hands to signify that he wanted to be chased.

While I did not expect Nim to invent signs, I felt that he should be encouraged to do so. The invention of signs seemed to provide strong evidence of Nim's interest in using signs to communicate. Mainly for that reason, I thought it wise to depart occasionally from our practice of requiring Nim to follow the topography of signs, as used in ASL, as closely as possible. Of course, a major advantage of requiring Nim to sign the signs of standard ASL was that a person not familiar with Nim but fluent in ASL might be able to read his signs directly. Also, if Nim were to meet another chimpanzee who knew ASL, it would be desirable for both chimpanzees to sign in the same language. In the case of *play*, however, members of the project agreed that it would be reasonable to make an exception and accept Nim's sign, particularly since we hadn't yet tried to teach Nim the standard sign for *play*.

Top: Nim invented his own sign for *hand cream*. Center: He recognizes a picture of hand cream and names it. Bottom: He enjoyed applying hand cream to his teachers' arms.

Nim's use of *play* was extensive. When he met a new child, he frequently signed *play*. In most cases the child was so thrilled by the opportunity to play with a chimpanzee that he or she readily obliged by chasing Nim. Nim's use of *play* was by no means limited to humans. Particularly with dogs, who were likely to chase Nim without any invitation, Nim frequently signed *play*. Nim also signed *play* to cats and horses. As far as I could determine, Nim never signed *play* to a picture or to an inanimate object. When looking at a picture book of animals or seeing a toy animal, Nim often signed *cat*, *dog*, or *bird*. *Play* seemed to be used exclusively with animate beings.

At various stages of the project there was some evidence that Nim may have tried to use other signs he invented. Teachers often reported that Nim persisted in making a new sign, often in a particular context. These signs were referred to as X signs and were discussed at weekly meetings. In some instances it appeared as if Nim was playing with his hands and not actually signing. In others, it seemed reasonable to believe that he may have been trying to express something for which he did not have a sign. Where there was enough agreement that Nim was attempting to communicate about something via an X sign, all of Nim's teachers were instructed to watch for that sign and try to deduce its meaning. Unfortunately none of the X signs materialized as additions to

Another sign he invented himself: *play*

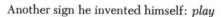

Nim's vocabulary. Just the same, I think it is reasonable to regard them as bona fide attempts on Nim's part to communicate. I believe that *X* signs did not satisfy our criteria for including a sign in Nim's vocabulary mainly because it was difficult to obtain a consistent response to an *X* sign. *Play* was used in a high-motivation situation that required no additional objects and was reinforced immediately by a number of teachers. Since Nim's *X* signs occurred in situations that were repeated less frequently, his feedback was probably insufficient to maintain the *X* signs that he tried to use.

Three of Nim's signs for colors, *green*, *yellow*, and *blue*, were modifications of standard ASL signs. They differed from other color signs that Nim had learned (*red*, *orange*, *black*, and *brown*) in one important respect. *Green*, *yellow*, and *blue* were each signed by moving the right hand, in a different configuration for each color, in a clockwise circle. The configuration is specified by the first letter of each color. *Blue* is signed holding the *b*-hand (see Appendix A) off to the side of the body and moving that hand configuration in a clockwise circle. *Yellow* and *green* are signed in the same manner, except they are made with the *y*-hand and the *g*-hand respectively. *Red*, *black*, *orange*, and *brown* are signed differently. In each case, a different hand configuration touches a different part of the body.

It was not that difficult to mold Nim's hands into the correct configurations for the color signs *green*, *yellow*, and *blue*, but it was difficult to have Nim make a sign without touching another part of his body. Try as we would to mold the configurations required for each of these colors into the clockwise motions specified by standard ASL, we made no progress. Accordingly, we required that Nim touch his stomach with the hand configuration appropriate for each color. We chose the stomach as a locus for the signs *green*, *yellow*, and *blue* because it was not involved in many of Nim's other signs. Once we gave Nim a part of his body to touch with his signing hand, it was not difficult for him to sign *green*, *yellow*, and *blue*.

In general, signs in which a hand makes contact with another part of the body (so-called "contact" signs) were much easier to learn than signs that entailed no contact. For example, the sign *yogurt*, which is the same as the ASL sign *milk*, is made by holding the right hand in a fist configuration off to one side of the body as if milking a cow. Nim required unusually long periods of training before he mastered *yogurt* and other signs that entailed little or no contact between the hands and another part of the body (for example, *box*, *hello*, and *finished*). Studies of sign-language-learning by deaf children also report difficulty with noncontact signs.

In the case of some signs, we accepted approximations of standard

Nim making his own versions of the signs for *blue* and *yellow*. Left:
Nim: *blue* / Dick: *yes*. Right: Nim: *yellow* / Bill: *good*

ASL signs that were referred to as "baby signs." Through the concerted
efforts of his teachers, Nim was slowly weaned away from the baby
configuration toward the adult version of a sign. A similar development
has been observed in children who learn sign language as a first language.

A transition from the baby to the adult version of a sign was just one
of a number of variations from human sign language that Nim was observed to make. In order to emphasize a particular sign, Nim often made
the sign with both hands. In normal practice only one hand is required.
We have seen Nim signing *dirty* with both hands to emphasize his need
to use the toilet. He would also use both hands to underscore a request
for a favorite food.

Errors provide yet another interesting example of systematic variation in Nim's signing. Our meager data on these errors also point to some
interesting similarities between sign language as practiced by humans
and by Nim. The significance of these errors is most easily appreciated
by considering how humans remember a list of unrelated words: for
example, *bad*, *cat*, and *big*. The errors that occur in this process are
phonetic and not semantic. They have more to do with the sounds of the
words than with their meanings. In attempting to recall lists, we do not
substitute words like *rotten* or *wicked* for *bad*, *pet* or *feline* for *cat*, or
large or *huge* for *big*. Instead, words like *pad* are substituted for *bad*,
cap for *cat*, and *pig* for *big*. Phonetic errors in list-learning of spoken

These three pictures show Nim signing *more* at 2 (with Amy), at 2½ (with
Laura), and at 3½ (with Bill). At first, he touched only his index fingers
together. Later, he touched the index fingers and the remaining fingers, but in
separate groups. Eventually, he learned to sign the standard form of *more*.

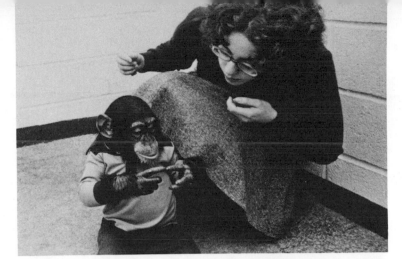

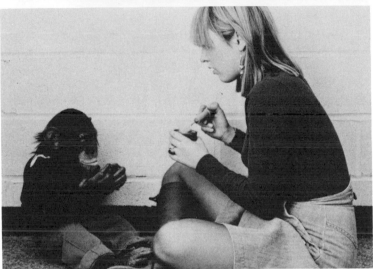

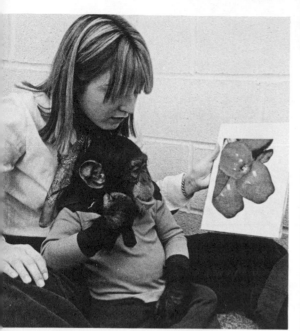
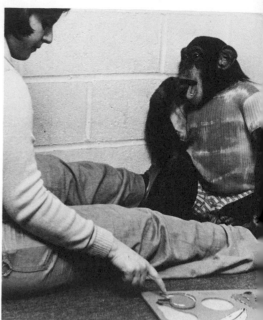

A two-handed *apple* sign. Joyce had not responded to Nim's original request for some apple.

words have an analog in the list-learning of signs. Signs that are made in a similar fashion are often substituted for one another. In learning to make my name sign, Nim often signed variations of *cat*, a sign that is topographically similar to *Herb*.

Nim's knowledge of signs was revealed most dramatically when he signed spontaneously in recognition of an object, person, or attribute. While driving Nim home one day, I noticed that he was signing *drink* repeatedly while we were waiting for a traffic light to change. I doubted that Nim was thirsty, since I had given him a bottle of orangeade to drink just before we left his classroom. A few seconds later, Nim pointed to a bus that had stopped alongside our car. The bus driver had just poured himself a cup of coffee from a thermos mug and was taking a sip while waiting for the light to change, and apparently this was what Nim was signing about. A few moments later, I confirmed that Nim wasn't thirsty by offering him a variety of drinks in the kitchen at Delafield.

I can recall a number of times when Nim described things that attracted his attention: signing *hat* while looking at a large picture on a billboard of a cowboy wearing a hat, and signing *red* while walking past a red flower. Most of his spontaneous signing occurred when he wanted something to eat, drink, or play with, but it is significant that

Signing *apple* in various contexts

Nim confused the similar signs *Herb* and *cat*. A correct *Herb* sign is shown at lower right; a correct *cat* sign will be seen on page 173 (bottom left). At top left, Nim wants to sign *cat* but comes up with a one-handed *cat* sign combined with an approximation of *Herb*. At top right, a one-handed *cat* sign is used to mean *Herb*. At lower left, Nim signs *cat* with one hand and *Herb* with the other, where *Herb* would be appropriate.

many of these requests occurred in the absence of the object that Nim signed about. He might sign *eat banana* when no banana was in sight; requests such as *hand cream, tea, drink, apple,* and *cat* also occurred without the desired object in sight.

As is true with a child, Nim understood more signs than he expressed. But it is easier to evaluate what words a child or chimpanzee can express than those it understands. When evaluating expression, it is necessary only to observe whether a particular sign occurs and in what context. In assessing comprehension, it is necessary to devise behavioral tasks that will show that understanding is specific to the sign and not to some other cue that the teacher may be transmitting. If the teacher signs *no* to get Nim to stop what he is doing, Nim may stop, not because he understands *no*, but because the teacher is also shaking his head. In evaluating comprehension, it is also necessary to ask from how many other signs can the sign in question be distinguished. Consider Nim's understanding of the sign *red*. It is possible that he understood only the difference between *red* and one other color sign, say *black*. That is, given a set of red and black objects, Nim might pick up a red object when his teachers signed *red* and a black object when his teacher signed *black*. That demonstration is sufficient to show that Nim understood *red*. But the breadth of his understanding would be much greater if he could differentiate between *red* and the signs for many other colors (for example, *blue*, *yellow*, *green*, and *brown*).

In many instances our basis for concluding that Nim could comprehend a sign came from tests performed in the classroom. For example, his teacher would arrange Nim's brush, a bottle of hand cream, a mirror, and other grooming articles on the floor. Nim would be positioned beside his teacher, equidistant from each item. The teacher would sign, *Nim, you give me hand cream*, deliberately not looking at or pointing to the object in question. Nim would reliably walk across the room, get the hand cream, and bring it to the teacher. If the teacher signed, *Nim, you give me brush*, Nim would walk over, pick up the brush, and place it next to the teacher.

Another variation of this procedure was to hold a closed but familiar book in front of Nim. The teacher would then sign, *Nim, where banana?* or *Show me banana*. Nim would often respond by looking through the book, finding the picture of the banana, and placing the book with the page open to the banana in front of his teacher. He would then sign *banana* and point to it. On one occasion, Nim was in the hallway outside his classroom when Laura signed *toys*. He interrupted what he was doing, ran into the classroom, and began taking the toys out of his toy bag. This sign was made in a totally new situation, but it nevertheless evoked the appropriate behavior.

One example of Nim's receptive vocabulary proved very fortunate. Most dangerous household products were locked in a cabinet in the kitchen. On one occasion it was left open, a fact that did not go unnoticed by Nim. While standing by the sink, Laura caught a glimpse of Nim at the other end of the room about to drink an open jar of rug cleaner. She signed frantically, *No stop don't eat*. Nim took the jar away from his mouth and put it down beside him.

While there is no limit to refining tests of comprehension, we felt that the tests we used adequately demonstrated Nim's responsiveness to specific signs. In each case, his behavior was both appropriate and immediate. In many of our tests, alternative responses were possible. For example, when Laura signed *toys* to Nim, there were many desirable objects nearby besides the toy bag, including his stroller, games, books, and a sink. Despite these other choices, each of which occasioned different behaviors, Nim responded by going directly to the toy bag.

A list of signs that Nim comprehended, according to tests administered independently by at least two of his teachers, appears in Table 3. Nim's comprehension of signs made it possible to engage him in conversation with his teachers about many topics. His teachers would ask him questions such as *What color?*, *What name of?*, *Who?*, and *Run?* Nim showed his comprehension by making an appropriate response.

Table 3.
Signs Nim Comprehended

afraid	car	first	Joyce
airplane	cat	fish	jump
alone	chair	flower	key
Andrea	change	fruit	kiss
angry	clean	give	later
apple	climb	go	Laura
attention	close	good	lie down
baby	coat	goodby	light
bad	color	grape	listen
ball	come	green	little
balloon	cookie	groom	locative
banana	crayon	gum	look
belt	cup	gym	make
berry	cut	hand cream	Mary
big	diaper	handkerchief	match
Bill	Dick	happy	me
bird	dirty	harmonica	mine
bite	dog	harness	mirror
black	don't	hat	more
blue	door	hello	mouse
Bob	down	help	mouth
book	draw	Herb	music
bowl	drink	here	napkin
box	ear	hot	Nim
break	easy	house	no
bring	eat	hug	nose
brown	egg	hungry	now
brush	eye	hurry	nut
bug	fall	hurt	on
butterfly	false	ice	one
camera	finish	in	open

orange	red	Steve	tree
out	Renee	stop	under
paint	right	Susan	up
pants	rock	sweet	wagon
paper	run	swing	wait
peach	shirt	table	walk
pear	shoe	take out	Walter
peekaboo	sign	taste	want
plant	sit	tea	wash
play	sleep	teeth	water
play key	smell	telephone	what
polc	smile	thirsty	where
pool	sock	throw	who
pour	sorry	tickle	window
powder	spaghetti	time	with
pull	spoon	toilet	work
put-in	squirrel	toothbrush	yellow
quiet	stand up	toys	yes
raisin	stay	train	you

As his ability to sign improved, Nim began to reply to his teachers' questions with more than one sign. Quite spontaneously Nim began to combine signs. During a two-year period, his teachers recorded almost 20,000 instances of combinations of two or more signs. It was this corpus of combinations that enabled me to try to answer the question that had led me to start Project Nim in the first place: can a chimpanzee create a sentence?

11

Can a Chimpanzee Create a Sentence?

The moment a child utters its first word is a moment of joy for its parents. In their view, the child has taken its first step on a familiar path toward the mastery of a language. So taken for granted is this process that few parents make as much of a fuss when their child utters its first sentence. Even though the essence of human language lies in the ability to create sentences, parents have come to assume that this ability follows naturally and automatically once their child learns to express itself in words.

Much more than parents, psychologists and psycholinguists have concerned themselves with the nature of a child's first sentences. They have tried to understand the precursors of sentences and how a child's early sentences differ from those of an adult. By examining the nature of a child's language, psycholinguists also hope to understand how language evolves from "baby talk" to conventional language. Following the practice of other developmental psychologists, psycholinguists often work from the simple to the complex. They assume that by studying the circumstances under which a child says "Toy Billy," it should be possible to obtain insight into the circumstances under which the same child might at a later stage of life say, "I'm going to give the toy to Billy."

Whatever the nature of the stages through which a child progresses in its mastery of a language, a psycholinguist can assume safely that at some point the child will begin to utter sentences. Obviously, this assumption cannot be made when studying language development in the chimpanzee. While chimpanzees have learned sizable vocabularies and have been observed to combine the words of these vocabularies into extended utterances, it is still not clear whether such utterances qualify as sentences.

Whether a chimpanzee can create a sentence would be easy to determine if there were a standard test for a sentence. Unfortunately, none exists. Ironically, one of the main reasons that this question has been given a lot of recent attention is that, for the first time, there is a compelling basis for asking whether a nonhuman species can create a sentence. Humans have no difficulty judging whether human utterances are sentences, but the basis for such judgments remains elusive. Intuitively, it seems straightforward to regard a sentence as a complete semantic proposition that is expressed through a set of words and phrases, each bearing particular grammatical relations to one another (e.g., actor, action, object). Such definitions may clarify the content of the sentence. They do not, however, provide a concrete basis for making judgments as to whether a particular string of words qualifies as a sentence.

At present, the basis for deciding whether a sequence of words deserves recognition as a sentence is an unwieldy process of elimination. When considering sequences of words, a linguist superimposes different kinds of "filters" on those sequences. By testing simpler explanations of the sequences under study, the linguist decides whether it is reasonable to invoke a grammatical rule to account for their occurrence. Thus, in deciding whether a chimpanzee has created a sentence, we must first eliminate simpler explanations of the evidence at hand.

In Chapter 2, I explained why a given sequence of words is not necessarily a sentence: it might have been learned by rote, or it might simply be a string of unrelated words. A child who learns to recite a prayer or a sentence such as "Oh say can you see?" often does so without any understanding of the relationships that exist between the words of the sequence he uttered. Accordingly, that child would not be expected to produce related phrases in different contexts, for example, "Oh write can John drive?" or "Oh can't you tell if Bill is there?"

There is no reason to regard sequences learned by rote as sentences generated by the application of grammatical rules. Chimpanzees are capable of memorizing dozens of sequences, such as *Please machine give apple, Please machine show slide,* and *Please Tim give Coke.* But unless it can be demonstrated that a chimpanzee, or a child, can create new sentences of the same structure that are appropriate to new contexts, their memorized sequences cannot be considered evidence of an ability to create sentences.

Even when it can be shown that a sequence of words was not learned by rote, it does not qualify automatically as a sentence. The words of a sequence may be relevant to a particular situation without being related to one another. Recall, for example, Washoe's sequence *water bird,* which she was reported to have uttered upon seeing a swan (see Chapter 2). We have no way of knowing whether she was signing that she saw a bird whose habitat was water or whether she was making two statements: "that is a bird" and "that is a body of water." A human would interpret "water bird" as a phrase in which "water" functions as an adjective to qualify "bird." But it doesn't follow that a chimpanzee would use the words in the same way. Unless we have evidence that the chimpanzee had a general rule for pairing adjectives with nouns (either as adjective-noun or as noun-adjective constructions), it is more reasonable to interpret their spontaneous sequences as a series of separate words related to a particular situation but not to each other.

Before I could argue that a chimpanzee can create a sentence, it was clear that I would first have to defeat the hypothesis that the chimpanzee's sequences were the product of rote learning or that they were just strings of unrelated signs. Showing that Nim's utterances were not rote sequences was much easier than showing that the signs of a particular sequence were grammatically related. By not requiring Nim to utter sequences of signs and by not requiring him to sign particular sequences, it was a relatively simple matter to rule out rote learning as an explanation of his sequences. Of course, one could argue that Nim modeled his sequences after those uttered by his teachers. But so long as there was no pressure to imitate precisely what Nim's teachers signed and so long as Nim's teachers did not correct his sequences of signs for grammatical mistakes, there is no basis for interpreting his signing as the product of rote learning. Unlike Sarah's teachers and Lana's computer, Nim's teachers did not differentiate between *Give Nim banana*, *Banana give Nim*, *Banana Nim give*, and so on. The impetus for Nim to model his utterances after those of his teachers seemed similar to that which a child experiences to imitate what its parents say.

In order to decide whether the signs of Nim's sequences were grammatically related to one another, it was necessary to obtain a large body or "corpus" of spontaneous utterances. Without such a corpus it would be impossible to decide whether a single utterance such as *water bird* was an interesting anecdote that, by chance, followed English word order or whether it was a manifestation of a rule for combining adjectives and nouns. In evaluating the utterances of Nim's corpus, it is important to keep in mind that Nim was not required to sign more than a single sign. If Nim wanted to be tickled or if he wanted to tickle his teacher, all that he had to sign was *tickle*. If he wanted to communicate that he wanted to be tickled again, either *tickle* or *more* was acceptable. Nim could sign *me* if he wanted to indicate that he and not the teacher was to be tickled. Before tickling his teacher, Nim could sign *you*.

Now suppose that Nim signed *you tickle me* as a request to be tickled. Should this utterance be interpreted as a sentence in which the meaning of one sign is modified by its combination with one or more additional signs or as a sequence of separate signs, each appropriate to the context at hand? If *you tickle me* was our only example of a sequence of signs, it would amount to nothing more than an interesting anecdote, interesting because it matched the order of words used to make a similar request in English. The sequence *you tickle me* could not, however, be regarded as evidence that Nim could create a sentence. That conclusion would be valid only if, at the very least, Nim favored the sequence *you tickle me* at the expense of the five other ways of combining the signs *you, tickle,* and *me: you me tickle, me you tickle, tickle you me, tickle me you,* and *me tickle you.*

In considering the number of sequences that might be generated by combining two or more signs, notice that the number of possible sequences increases rapidly as the number of signs increases. With two signs, there are only 2 possible sequences. With three, there are 6 possible sequences. With four signs, it is possible to generate 24 different sequences, with five sign, 120 different sequences, and so on.

Let us again consider the sequence *you tickle me.* Suppose that Nim used this sequence reliably when he wanted to be tickled. Is it then reasonable to conclude that Nim created a sentence by arranging the signs *you, tickle,* and *me* in what a human would regard as a subject-verb-object sequence? Clearly not. Nim may have used the sequence *you tickle me* because he preferred that sequence. When the number of alternative sequences is fairly small, a chimpanzee's memory would allow it to associate a small number of preferred sequences with specific situations such as tickling, playing, being fed, and so on. But as good as a chimpanzee's memory may be, it would undergo greater and greater strain in producing particular sequences as the number of possible alternatives increased. It would also become more and more difficult to explain a particular sequence by saying that the chimpanzee somehow memorized it, if it could be shown that the chimpanzee could substitute appropriate signs in each position of the sequence.

An important step in arguing that Nim generated the sequence *you tickle me* by using a rule such as "place the subject in the first position, the verb in the second position, and the object in the third position" is to show that Nim could substitute other signs within each position of the sequence. For example, one would want to see whether Nim would sign sequences such as *you tickle Nim* (when Nim wanted to be tickled), *Herb tickle me* (when Nim wanted to be tickled by Herb), or *Nim tickle Bill* (when Nim was about to tickle Bill). Alternatively, Nim might sign *you hug me* (when he wanted to be hugged), *Susan hug Nim* (when he wanted to be hugged by Susan), or *Nim hug cat* or *me hug cat* (when he wanted to hug another creature).

Only by showing that Nim could substitute appropriate signs in each position of a sequence and that it was unlikely that the substitutions were memorized could one have a basis for concluding that a sequence was generated by a rule. It is in this sense that linguists have argued that unlike words, which can be memorized one at a time, sentences are not learned individually.

From June 1, 1975, through February 13, 1977, Nim's teachers observed him signing more than 19,000 utterances that contained two or more signs. The basic method used to obtain this data was simple. During each session, Nim's teacher whispered vital data about his signing into a miniature cassette recorder: what he signed; with which hand; whether his utterance was molded, prompted, spontaneous, or imitative; and as much of the context of the utterance that could be described succinctly. After the session, his teacher transcribed the tape and prepared a detailed report on Nim's utterances.

Some sessions were videotaped and the videotapes were transcribed. Videotape transcripts were valuable for a number of reasons. The time pressure of having to dictate what Nim signed and also to dictate whatever could be grasped about the context of this utterance made it difficult to provide detailed descriptions of the context of each utterance. A videotape provided a permanent, and therefore much more leisurely, basis for characterizing the context of Nim's signs.

One important aspect of the context of Nim's utterances that was difficult to record by dictating into a miniature cassette recorder was what the teacher was doing and signing before Nim signed. Not only was there too much else to record, but as we shall see in Chapter 13, Nim's teachers were often unaware of what they had signed to him. On videotape, however, teachers' signs and their contexts were readily discernible. Videotape also provided an especially valuable opportunity to check the reliability of teachers' reports. To the extent that the data obtained from the videotapes agreed with that obtained from the reports, we could be confident of their accuracy. I am happy to say that we obtained high measures of reliability from these and from other checks.

The combinations included in our corpus had to satisfy a number of rules. These rules insured a conservative and unambiguous picture of Nim's signing. Our most basic rule defined a combination of signs: the occurrence of two or more signs that were not interrupted by the return of the hands to a resting or relaxed position or by any other action. The segmentation of signs into combinations is similar to the segmentation of speech into phrases. In both instances, the rules of segmentation group together words that are immediately related to one another.

Identifying a sequence of signs as a combination did not guarantee its inclusion in the corpus. Because of the spatial nature of sign language, it was not always possible to tell the order of signs in combination. Accordingly, we adopted a set of rules that excluded so-called nonlinear combinations from our corpus. These rules insured that the corpus would consist entirely of combinations in which signs occurred unambiguously one after the other, i.e. linear sequences.

We excluded from the corpus all contractions and simultaneous combinations. In spoken language, contractions (such as "can't" and "isn't") are commonplace. However, the scope of contractions in spoken language is much more limited than in sign language. Contractions in spoken language consist mainly of verbs and negatives, verbs and auxiliaries ("wanna"), and verbs and pronouns ("gimme," "woncha"). Thanks to the spatial property of sign language, it is possible to contract a much larger variety of signs. An example of the contraction of signs can be seen on page 247: the signs for *past* and *week* are combined to form the contraction *last week*. This type of contraction is not found in spoken language.

Linear sequences accounted for approximately 95 percent of Nim's combinations. The remaining 5 percent were combinations in which two or more signs occurred at the same time: simultaneous combinations or contractions. In our corpus we counted only linear sequences, where there was no ambiguity as to the order of the signs that were combined.

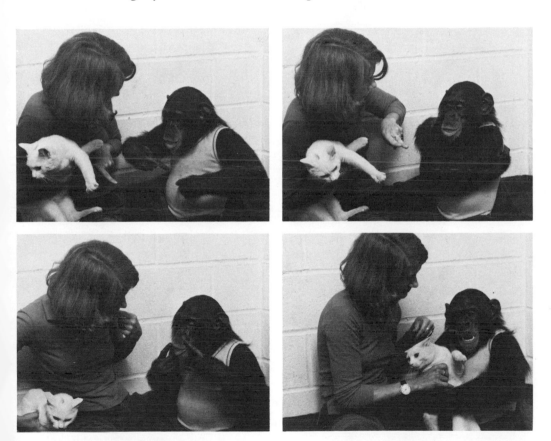

A linear sequence. Starting at top left: *me . . . hug . . . cat.* Lower right: the reward.

Another linear sequence: *tree . . . there*

Two signs combined: *Nim* (compare page 149) and *hug* (compare page 173, top right)

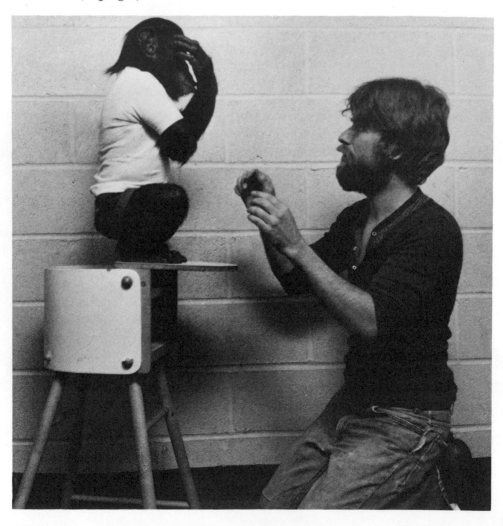

The sequence of signs *me $\genfrac{}{}{0pt}{}{more}{eat}$ banana*, illustrated in the series of photographs shown below, poses an interesting problem of interpretation. The signing in the third picture of the series seems to be a contraction. Nim has just been given a banana for signing *banana*. Given that he is signing *more* with the same hand configuration used to sign *eat*, and given other examples of contractions involving *more* (for example, *more drink*—see the *tea* sequence on page 141), it seems reasonable to interpret the combination as a contraction of *more* and *eat*. An alternative explanation is suggested by the picture on page 163, in which Nim is shown making a two-handed *apple* sign. Why not interpret Nim's signing as a two-handed *eat* sign? I decided against this because it did not seem to convey the emphasis of other double-handed signs, which were usually repeated a number of times and were signed with vehemence. They often

Nim signs *banana* and receives the appropriate reward (upper left). But he is not satisfied. He signs *me* (upper right) . . . *more/eat* (contraction; lower left) . . . *banana* (lower right).

occurred when the desired object had been withheld longer than usual. That seemed to be the reason for the double-handed *apple* sign. Clearly, my interpretation cannot be accepted as conclusive. But it does reveal some of the interesting ambiguities avoided by including only linear combinations in the corpus.

There was no ambiguity in Nim's simultaneous combination of *me* and *hat*. Both *me* and *hat* were signed exactly as they would have been if they had occurred sequentially. But this is a relatively simple example of Nim's simultaneous signing. More complicated examples arose when Nim maintained a particular sign with one hand while signing a sequence of other signs with his other hand. Consider the following excerpt from a video transcript of Nim asking for a grape and a sip of tea.

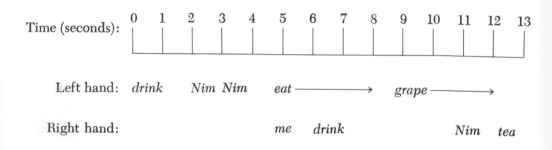

Here Nim maintained the sign *eat* with his left hand while signing *me drink* with his right hand. Subsequently he signed *grape* with his left hand and continued doing so while signing *Nim* with his right hand.

Two rules were used to insure the shortest possible description of a particular combination in the tabulation of combinations containing successive repetitions of the same signs.

Homogeneous sequences: Consider a combination such as *eat eat eat*, in which all the signs are the same. Such combinations were treated as a single-sign utterance. Homogeneous sequences of signs were not tabulated as combinations.

Heterogeneous sequences: Consider the combination *banana me me me eat*. This combination consists of three different signs with one sign repeated successively. Before it was entered in the corpus, a sequence of heterogeneous signs containing the successive repetition of one or more signs was reduced by deleting all repetitions. Thus *banana me me me eat* was reduced to *banana me eat*. Whereas the original sequence contained five signs, this combination was entered as a three-sign sequence. But notice that this rule applied only to combinations in which a sign was

Herb: *mine* / Nim: *me*/*hat* (simultaneous signs)

repeated successively. A combination such as *me banana me eat me* was not reduced.*

The corpus of linear sequences was tabulated in two ways. One tabulation considered all instances of linear sequences. The other counted only the first instance of a particular multisign sequence. Each distinct multisign sequence is a combination "type"; each instance of a particular type is a "token" of that type. In tabulating tokens an equal weight was attached to the occurrence of any utterance, whether or not it had

---

* In ASL, repetitions of a sign often create a noun from a verb. Many verbs (for example, *sweep*, *fly*, and *drive*) are made with a single motion. Related nouns (*broom*, *airplane*, and *car*) are made by repeating those signs twice (the so-called "double bounce" form). None of Nim's teachers could discern any particular meaning in Nim's repetitions of signs. Certainly he was not distinguishing between nouns and verbs, and we saw no evidence that Nim's repeated signs were a kind of "stuttering." The sole reason for Nim's repeating a sign seemed to be perseverance. Overall, fewer than 10 percent of Nim's linear utterances contained successively repeated signs, too few to make any difference in the outcome of our analysis of Nim's combinations.

occurred previously. For example, in totaling the number of tokens, each of the utterances *tickle me, tickle me, tickle me* would count once. In totaling the number of types, only the first occurrence of *tickle me* counted. However, each time Nim used the two signs in a different order (as in *me tickle*), that counted as another type.

An overall view of Nim's tendency to form combinations of signs during the period June 1, 1975 to February 13, 1977 can be seen in these

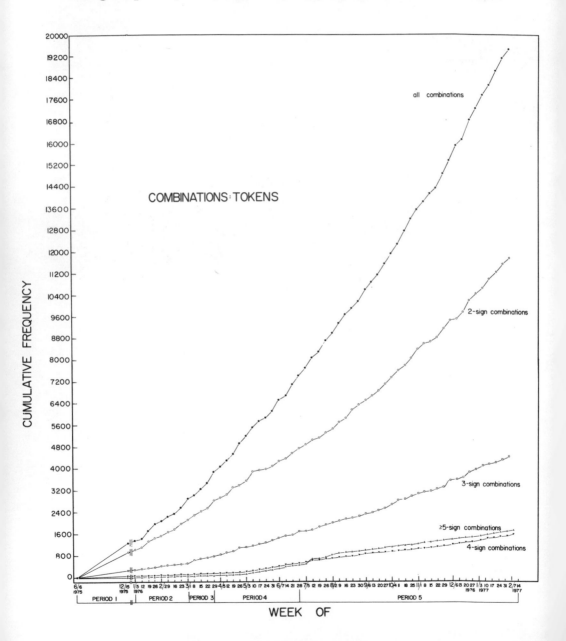

graphs, which show the cumulative frequencies of tokens and types of combinations of two, three, four, five, or more signs. Nim's first combinations (*more drink* and *more eat*) occurred on March 3, 1975, the day after he satisfied the acquisition criteria for his first sign. Since then, he has made numerous combinations, some containing as many as sixteen signs. Overall, there were 19,203 tokens of 5,235 types of combinations of two, three, four, five, or more signs.

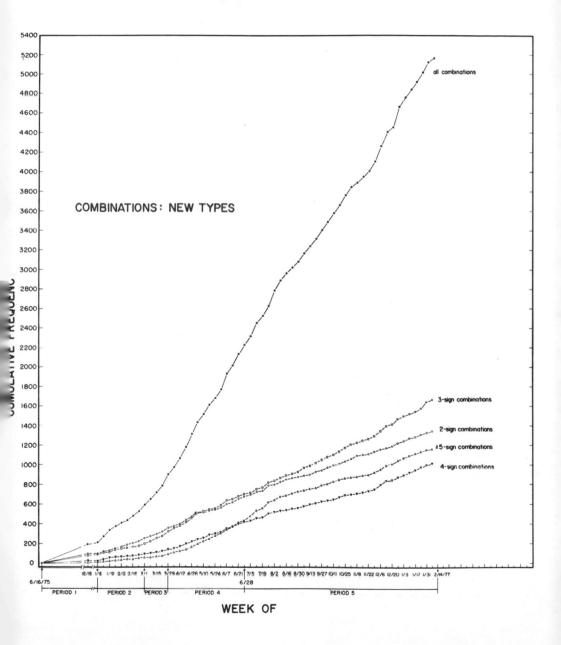

The sheer variety of types of combinations and the fact that Nim was not required to combine signs suffices to show that his combinations were not learned by rote. Considering only Nim's two- and three-sign combinations, the ocurrence of 2,691 types of combinations would strain the capacity of any known estimate of a chimpanzee's memory. As mentioned earlier, however, a large variety of combinations is not sufficient to demonstrate that such combinations are sentences.

Two characteristics of the corpus we collected suggest that some sort of structural rule is needed to account for the variety and number of Nim's combinations. One is the regularity of his combinations. These regularities are most easily seen in the case of two-sign combinations. For example, sequences of the form *more* + X occurred with greater frequency than sequences of the form X + *more*; and *verb* + X constructions were more frequent than X + *verb* constructions; and so on.

The analysis that showed that certain kinds of constructions were more frequent than others did not interpret the meanings of those combinations. The frequencies of different types of combinations were simply noted and compared with one another. Because this type of analysis considers only the distributions of signs in combination with one another, it is called a "distributional" analysis. A second line of evidence supporting a structural interpretation of Nim's combinations emerges from semantic analyses of these combinations. Unlike the distributional analysis of our corpus, which simply tabulated the number of tokens of each type of combination, our semantic interpretations were based on an analysis of Nim's actions and intentions, and their content.

Consider first the regularities that emerged from a distributional analysis, which is presented in Tables 4-9 on pages 185-91, of Nim's two-sign combination. Table 4 shows all two-sign combinations of the corpus containing *more*. Clearly, there were considerably more types and tokens containing *more* in the first than in the second position. This was true whether *more* was combined with signs glossed as nouns (such as *banana*) or as verbs (such as *tickle*). A similar state of affairs can be seen in Table 5, which shows all two-sign combinations containing *give*. Here again there is a strong tendency for *give* to occur in the first position.

In the case of combinations containing *more*, it might be argued that Nim modeled the construction *more* + X after his teachers' utterances. Often a teacher would sign *more* + X? to Nim to see if he would sign *more*, X, or a related sign in reply. In some instances Nim signed *more* + X even though he wasn't required to do so. In this view, he learned to sign *more* + X, first by imitating a few instances of *more* + X and then generalizing this construction to new objects and actions. This type of explanation seems less cogent in the case of *give* + X. Nim began signing *give* + X reliably long before his teachers asked him to give them objects by signing *give* + *X*. Nim was much too self-centered when he began to sign *give* + *X* to provide any hope that he would voluntarily give up an object his teacher might ask for.

Nim's use of transitive verbs and references to himself, either as *me* or *Nim*, provides another interesting example of regularities in Nim's two-sign combinations. Table 6 shows Nim's two-sign combinations containing transitive verbs such as *hug, chase,* and *tickle* and either *me* or *Nim*. Both the number and type of tokens with the verb in the first position far exceed the reverse kind of construction. Imitation of his teacher's signing is again an unlikely explanation. Nim was signing *tickle me* long before his teachers asked him to tickle them by signing *tickle me* or *tickle* + teacher's name sign.

Table 6 also shows that Nim combined transitive verbs as readily with *Nim* as with *me*. Virtually the same number of types of sequences contain *Nim* as do *me*. That there were more tokens of two-sign verb combinations containing *me* than *Nim* is perhaps best explained by the fact that Nim learned to sign *me* before he learned to sign *Nim*. During the most recent period of this analysis (July 5, 1976 to February 13, 1977), the frequency with which *Nim* and *me* were combined with transitive verbs was essentially the same.

Nim's preference for using *me* and *Nim* in the second position of two-sign combinations was not restricted to combinations containing transitive verbs. This preference was also apparent, though to a lesser extent, in combinations containing the names of food and drink items. Table 7 shows all two-sign combinations containing *me* and *Nim* combined with food or drink nouns. A somewhat lesser preference for the location of the signs *me* and *Nim* is apparent in the case of two-sign combinations in which these signs were combined with nonfood/drink nouns (Tables 7 and 8). The percentage of combinations in which *me* and *Nim* appear in the second position was highest when those signs were combined with transitive verbs (83 percent), next highest when combined with food and drink nouns (75 percent), and lowest when combined with nonfood and nondrink nouns (65 percent). Why these differences in how Nim referred to himself? One explanation is that when Nim combined transitive verbs with food or drink nouns, he was using the signs *me* and *Nim* mainly as indirect objects. However, when Nim signed about objects that are neither edible nor drinkable, he may have signed *me* and *Nim* to indicate possession on some occasions, and to refer to himself as an indirect object on other occasions. For example, when Nim signed *hat me*, he may have been asking his teacher to give him the hat. But when he signed *me hat*, he may have been saying that he regarded the hat as his. These and other interpretations of Nim's signing will be considered below in our semantic analysis of Nim's two-sign combinations.

The absence of a universal pattern in combinations containing *me* and *Nim* suggests that Nim was not simply using a position habit to form combinations. That certain sign categories tended to appear more frequently in the first position (verbs and *more*) and others in the second (*eat* and *drink*) could simply mean that Nim differentiated the first and second positions of two-sign sequences irrespective of their meaning. Different frequency patterns involving signs like *me*, *Nim*, *give*, *more*, and transitive verbs may result entirely from independent "habits" of when to use particular signs.

How can the "position habit" hypothesis be tested in general? The hypothesis is that Nim tended to use certain categories as "initial" or "final," irrespective of the context in which they occurred. For example, Nim may have put verbs in the first position and *me* or *Nim* in the second position of two-sign sequences, not in order to create a specific meaning, but because that was his habit. If this was true, any regularities observed in Nim's combinations should be predictable on statistical rather than linguistic grounds. So long as one knows the overall probability of Nim's placing a sign in the first and second positions respectively, it should be possible to calculate how frequently combinations of particular signs would occur. The logic for making such calculations is the same as the logic that a card player follows in predicting that two successive draws from a well-shuffled deck will yield a heart followed by a club. That probability would be $13/52 \times 13/51 = 0.06$. If this type of draw is repeated many times, one would expect, on the average, a heart followed by a club to occur six times out of a hundred. That outcome is inherent in the organization of the deck and nothing else.

In order to evaluate the position habit hypothesis, we performed an extensive statistical analysis on the relative frequencies of different signs in each position of two- and three-sign combinations. The results showed that it was not possible to account for Nim's combinations by a rule that combined signs solely on the basis of their relative frequencies in each position of multisign combinations.

The sheer size of the corpus of Nim's combinations and the regularities observed allow us to dispense with the hypothesis that Nim was haphazardly combining whatever signs came to mind when he emitted a particular combination. But as gratifying as these results may be to those who believe that chimpanzees can learn to use language, it is not immediately clear what the regularities really mean.

It is important not to lose sight of the fact that the regularities resulted from nothing more profound than frequency counts of different types of combinations. The only meaning that can be inferred from these combinations is that which a speaking person projects onto them. But how do we know that these were the meanings Nim actually intended?

Distributional data concerning regularities of sign order can tell us that Nim's combinations were somehow constrained in the sense that they weren't random concatenations of signs appropriate to various contexts. But distributional data alone cannot tell us why Nim chose particular word orders. Such data cannot for example allow us to interpret *hat* and *me* as a request for a hat and *me* and *hat* as an assertion about possession. What is needed is evidence that Nim used sign order as a device for encoding certain meanings.

Consider the interpretation of a combination such as *Nim eat*. A distributional analysis considers what signs were made and in what order. A semantic analysis could render a number of different interpretations of the same signs. For example, *Nim eat* might have been uttered as a request for food. If the context of the sequence *Nim eat* supported that interpretation, it would be tabulated as an instance of a beneficiary-object combination. Other examples of beneficiary-object combinations are *Bill apple, Nim ball*. *Nim eat* could also mean that Nim was signing that he was eating. According to this interpretation, Nim is relating an agent to the action of eating. If the context of *Nim eat* suggested that interpretation, *Nim eat* would be tabulated as an instance of the action-agent category. Other examples of action-agent combinations include *Herb tickle, Joyce play*, and *Dick go*.

The best way to decide the meaning of a young child's utterance is to obtain as full a characterization of the context of that utterance as possible. For example, the child may reach for an object that he cannot quite grasp and say *Toy me*. His mother then gives him the toy, and the child smiles while clutching it. Given the child's behavior, his utterance, and his reaction to his mother's behavior, it is reasonable to interpret this utterance as describing a relationship between an object and a beneficiary. It is easy to imagine how other utterances might be interpreted as statements about relations between agents and actions (*Mommy give*), between an action and a beneficiary (*Give me*), between an attribute and an object (*red toy*), and so on.

Nim's teachers' reports did not contain enough information about context to provide a consistent basis for making semantic interpretations of his utterances, which is why we videotaped sessions both at Columbia and at Delafield. These tapes were transcribed by the teacher who taught the session and, as a reliability check, they were also transcribed by another teacher. In preparing their transcripts, Nim's teachers tried to interpret the meanings of as many two-sign combinations as possible. Eighty-four percent of the two-sign combinations obtained from this sample were assigned to one of eight semantic categories. These categories and examples of the types of utterances assigned to each category are shown in Table 9.

This data is gratifying for a number of reasons. For one thing, it was possible to interpret roughly the same percentage of Nim's utterances as is possible in studies of a child's utterances. Studies of child language at the stage when the child begins to make two-word utterances show that approximately 80 percent of those utterances can be interpreted as instances of about eight semantic categories, which coincide with those we used. Of major importance was the agreement that could be observed between the results of our semantic and distributional analyses. In the distributional analysis we observed that noun + *me* and noun + *Nim* constructions were more frequent than the reverse constructions. The semantic analysis showed that object + beneficiary constructions were more frequent than beneficiary + object constructions.

As far as I can tell, the distributional, statistical, and semantic analyses of Nim's utterances represent the most intense and systematic effort ever made to evaluate a chimpanzee's ability to create a sentence. At the very least, these data invalidate two simpler interpretations of word sequences emitted by a chimpanzee: that they were learned by rote or that they were random combinations of signs, each relevant to a particular context but unrelated to one another.

When we finally completed them during the summer of 1977, the tedious statistical analyses of Nim's signing behavior appeared to give us just what I had hoped for—a solid basis for demonstrating that a chimpanzee can create a sentence. This was a central goal of Project Nim, and the fact that we appeared to have achieved it successfully gave everyone's morale a boost. It was a happy reward for many weeks of double duty—days spent looking after Nim and evenings spent tabulating and analyzing mountains of data.

But scarcely another month had passed before I realized that my satisfaction may have been premature. Contrary to what I had thought, the statistical analyses were not definitive after all. Painstaking examination of videotapes of Nim's conversations with his teachers unexpectedly revealed an aspect of his signing behavior that I—and other experimenters—had hitherto overlooked. Ironically, the only reason I found the time to study the videotapes was a sad one: it was no longer possible to keep Project Nim going, and Nim himself had to be returned to Oklahoma.

Table 4.
Two-Sign Combinations Containing *More*

MORE + X		X + MORE	
Type	*Token*	*Type*	*Token*
more + apple	12	apple + more	5
		baby	1
ball	2		
banana	62	banana	5
berry	2		
bill	1		
bite	2		
brush	5		
chair	19	chair	3
drink	99	drink	14
eat	287	eat	58
fruit	2		
give	1	give	1
go	7	go	2
grape	11	grape	2
groom	4		
gum	29	gum	1
hand cream	23	hand cream	5
		hat	2
hug	16	hug	3
hurry	1	hurry	2
in	1		
jump	1		
key	1		
listen	1		
me	42	me	12
Nim	24	Nim	7
nut	11	nut	3
open	1		
orange	6		
paint	1		
peach	2		
pear	13		
play	41	play	7
pole	9	pole	1
raisin	1	raisin	1
		red	1

MORE + X		X + MORE	
Type	*Token*	*Type*	*Token*
shoe	2	shoe	1
smell	3		
spoon	2		
sweet	14	sweet	5
swing	1		
tea	23	tea	8
tickle	136	tickle	23
toothbrush	3	toothbrush	1
up	1		
water	10	water	1
what	6		
yogurt	5	yogurt	2
47 Types	974 Tokens	27 Types	174 Tokens

Table 5.
Two-Sign Combinations Containing *Give*

GIVE + X		X + GIVE	
Type	*Token*	*Type*	*Token*
give + apple	9	apple + give	3
baby	1	baby	1
ball	14	ball	1
banana	7		
black	1		
blue	2		
brown	1		
brush	3	brush	2
bug	2		
clean	1		
		come	3
		cracker	2
crayon	2		
dog	1		
drink	15	drink	7
eat	54	eat	12
finish	1	finish	3
flower	2		

GIVE + X		X + GIVE	
Type	*Token*	*Type*	*Token*
grape	3	grape	1
gum	4	gum	3
hand cream	14	hand cream	3
harmonica	2		
here	1		
hug	3	hug	1
hungry	2		
hurry	2		
jump	2		
key	1		
kiss	1		
light	2	light	2
listen	1		
locative	6	locative	2
me	41	me	11
more	3		
Nim	23	Nim	4
nut	2	nut	2
open	2	open	2
orange	3		
out	1		
pear	2	pear	2
play	1	play	3
point	2	point	2
raisin	2	raisin	2
red	2		
rock	1		
		shoe	1
smell	1		
spoon	1		
sweet	6		
tea	1		
tickle	1		
toothbrush	4		
water	9	water	4
what	1		
51 Types	271 Tokens	24 Types	77 Tokens

Table 6.
Two-Sign Combinations Containing *Me* or *Nim* and Transitive Verbs (V$_T$)

V$_T$ + ME		V$_T$ + NIM		ME + V$_T$		NIM + V$_T$	
Type	Token	Type	Token	Type	Token	Type	Token
bite + me	3	bite + Nim	2	me + bite	2	Nim + brush	4
break	2						
brush	35	brush	13	brush	9		
clean	2	clean	1	clean	2		
				cook	1		
		draw	1			finish	1
finish	1	finish	7			give	4
give	41	give	23	give	11	go	4
groom	21	groom	6			groom	1
help	6	help	4	help	2	hug	23
hug	74	hug	106	hug	40	kiss	2
kiss	1	kiss	6	kiss	1	open	5
open	13	open	6	open	10		
		pull	1				
tickle	316	tickle	107	tickle	20	tickle	16
12 Types	515 Tokens	13 Types	283 Tokens	10 Types	98 Tokens	9 Types	60 Tokens

Total Types: 25   Total Tokens: 788   Total Types: 19   Total Tokens: 158

Table 7.
Two-Sign Combinations of *Nim* or *Me* + Food/Drink Noun

NOUN + NIM		NOUN + ME		NIM + NOUN		ME + NOUN	
Type	+ Nim (Token)	Type	+ me (Token)	Type	Nim + (Token)	Type	me + (Token)
apple	65	apple	27	apple	25	apple	17
banana	73	banana	97	banana	18	banana	34
berry	1	berry	2				
cracker	21	cracker	3	cracker	3	cracker	1
egg	2	egg	2				
fruit	11	fruit	1	fruit	6		
grape	21	grape	12	grape	5	grape	2
gum	47	gum	19	gum	21	gum	43
nut	71	nut	16	nut	9	nut	4
orange	4						
pancake	2	pancake	2				
peach	3			peach	1	peach	1
pear	20	pear	4	pear	4		
raisin	23	raisin	5	raisin	6	raisin	4
sweet	85	sweet	23	sweet	13	sweet	8
tea	14	tea	17	tea	7	tea	13
water	10	water	13	water	2	water	5
yogurt	57	yogurt	2	yogurt	8	yogurt	1
18 Types	530 Tokens	16 Types	245 Tokens	14 Types	28 Tokens	12 Types	133 Tokens

Total Types: 34   Total Tokens: 775

Total Types: 26   Total Tokens: 261

Table 8. Two-Sign Combinations of *Nim* or *Me* + Non-Food/Drink Noun

NOUN + NIM

Type	+ Nim	Token
baby		20
ball		6
book		2
brush		13
bug		1
chair		2
hand cream		6
harmonica		1
hat		3
ice		2
key		1
pants		2
pole		1
shoe		3
smell		2
socks		1
spoon		3
toothbrush		17
18 Types		86 Tokens

NOUN + ME

Type	+ me	Token
baby		2
ball		7
brush		35
cat		1
chair		1
dog		2
hand cream		4
harmonica		1
hat		20
key		3
pants		4
pole		2
shoe		4
smell		1
spoon		1
toothbrush		6
16 Types		94 Tokens

Total Types: 34 Total Tokens: 180

NIM + NOUN

Type	Token
Nim + baby	6
bird	1
book	1
brush	4
bug	1
chair	2
color	2
hand cream	7
hat	8
key	1
music	1
pants	1
paper	1
shoe	1
toothbrush	4
15 Types	41 Tokens

ME + NOUN

Type	Token
me + ball	10
book	3
brush	9
flower	1
hand cream	3
hat	26
pants	2
pole	1
shoe	1
smell	1
toothbrush	1
11 Types	58 Tokens

Total Types: 26 Total Tokens: 99

Table 9.
Examples of Two-Sign Combinations in the Eight
Most Frequently Occurring Semantic Categories
(Frequency of Occurrence of Each Type)

ACTION + OBJECT (27%)
    eat grape
    drink tea
    out shoe

OBJECT + BENEFICIARY (16%)
    food Nim
    ice Nim
    baby Nim
    yogurt Nim

TWO PROPOSITIONS (16%)
    eat tickle
    dirty hug
    out open

ENTITY + PLACE (6%)
    baby chair
    food there
    grape there
    banana house

ROUTINE (6%)
    out pants
    in pants

ATTRIBUTE + ENTITY (5%)
    green color
    red apple
    orange balloon
    hungry me

ACTION + PLACE (5%)
    clean there
    out there
    tickle there

AGENT + ACTION (3%)
    Bill run
    Nim out
    Nim wash
    me open

12

Nim Leaves

Throughout Project Nim I was faced with the need to provide Nim with caretakers he liked, sixteen hours a day, seven days a week. This was the "baby-sitting" problem, and unless it was solved, no progress could be made in dealing with the raison d'être of the project: teaching Nim sign language and obtaining scientifically useful data about his knowledge of sign language. It all boiled down to a single problem: personnel.

By January 1977, Nim had experienced too many turnovers in the teaching staff, and I had not succeeded in raising enough money to attract the kind of teachers the project needed. Again, I had to ask myself a painful question that I had not been forced to confront since the prior summer: Was it scientifically defensible to continue the project if I could do little more than deal with the baby-sitting problem?

Another factor that made it difficult to continue Project Nim was the large amount of my own time it required. During the 1976–77 academic year I was on sabbatical leave, which I devoted fully to the project. After September 1, 1977, I would have to return to teaching and other duties at Columbia. Even if I wanted to continue living an existence in which my life was consumed by various aspects of the project, the amount of full-time days I would be able to devote to that task was numbered.

I was not the only member of the project who would have to cut back on his contribution to the project. Dick, Bill, and Joyce made clear that they could no longer continue their yeoman service with Nim and also fulfill their obligations as students once the 1977–78 school year began. But even if they could work full time, it was unclear how they would be paid. The funds I had available would run out long before August 31, 1978, the end of the two-year period of the NIMH grant.

Another long-standing problem was the steadily increasing volume of data we had collected. Most of it was unanalyzed. It was unsettling and downright frustrating to have worked on a project for almost four years without knowing precisely what we had discovered and without having published any results. But as much as I wanted to write about the project, I was aware that any effort in that direction would detract from the strength of the program for teaching Nim to sign. Data analysis and publication could wait, while lost time between Nim and his teachers could never be recovered.

At the beginning of July 1977, I decided that I would arrange to send Nim back to Oklahoma unless I could obtain at least three new teachers by September. During the summer I would also continue to look for the additional funds needed to pay those teachers. That effort to find new teachers turned out to be the largest I had undertaken since the start of the project. I wrote to and called colleagues in more than a dozen psychology departments in the New York area, and for the first time I advertised in a newspaper. Particularly because of the newspaper ad, the volume of responses was almost overwhelming. Unfortunately, the number of qualified applicants was disappointingly low.

At this stage of the project I knew more clearly than ever what kind of person would make a good teacher. I not only wanted the teacher to know sign language, but I also expected some teaching experience with children, an interest, or preferably experience, in doing research, and a mature and assertive personality. Not one of the applicants satisfied all of these requirements. Most disappointing was the lack of proficiency in sign language. Those applicants who seemed otherwise qualified were asked to improve their ability to sign (in classes I arranged) and also to observe and try to get to know Nim.

The trials of new people were hard on me and on the experienced teachers who supervised them. Worst of all, they were hard on Nim. He hadn't taken well to such trials previously, and he liked them even less during August 1977. By the end of August it had become painfully clear that the personnel I needed were not on hand and that there was no way to continue the project at the level we had thus far managed. At best it might be possible to arrange a baby-sitting service for Nim, which hardly justified a budget of more than $70,000 a year and all the headaches entailed in keeping that relatively simple but demanding operation intact. Indeed it was not even clear how effective a baby-sitting operation I could maintain. Nim's regular teachers (Bill, Joyce, Dick, and Susan) were working long and exhausting sessions. Whereas I was once able to schedule three sessions a day (each about three and a half hours long), it was now necessary to schedule two six-and-a-half-hour sessions a day. None of Nim's teachers could continue that back-breaking schedule after school began. There was no choice but to say *finis*.

During the third week in August I informed Dr. William Lemmon, the director of the Institute for Primate Studies, of Norman, Oklahoma, that I could not keep Nim beyond the end of September. I also informed Roger Fouts, of the University of Oklahoma, that Nim would soon become a resident at the institute. Roger, a student of the Gardners and

Washoe's main trainer, spent a lot of time at the institute doing research centered on signing with Washoe and other chimpanzees.

Despite my confidence in Dr. Lemmon and Roger Fouts, I had mixed feelings about returning Nim to the institute. Even though Nim would have the company of other chimpanzees, his existence there would mean confinement to a cage. One alternative I had considered was a chimpanzee reclamation center in Orlando, Florida. There, laboratory-reared chimpanzees were taught to cope with the junglelike vegetation of the Everglades. In discussing this and other possible new homes for Nim with Dr. Lemmon it became clear that he did not want to relinquish control of Nim. Since Dr. Lemmon had been kind enough to give me Nim as an infant, at no cost and for as long as I wanted, I felt obliged to comply with his wish to return Nim to the institute.

The target date for Nim's return was September 25. By that time I hoped that I would be able to complete some ongoing experiments and to arrange the complicated logistics of moving Nim to Oklahoma. The cheapest way to send him was in a crate, via air freight. I ruled out that alternative on the grounds that it would be too traumatic, especially since it would coincide with the sudden transition from Delafield to Oklahoma. The remaining options were to drive Nim to Oklahoma or to fly him there in a private plane. In either case Nim would have to be sedated for the first time.

Particularly because it was hard to predict how long a sedative would remain effective, I had little enthusiasm for driving Nim. Even if we managed to drive nonstop from New York to Oklahoma, I estimated that the trip would take at least twenty-four hours. At a cost that was considerably greater than I could afford, I chartered a twin-engine propeller plane and arranged to fly Nim to Oklahoma on September 24.

As a farewell to Nim I invited as many of his old teachers as I could locate in the New York area to come up to Delafield on September 23 to see him for what would most certainly be the last time and to pose for a group picture. The gray, drizzly day was suddenly brightened by the joy Nim showed when I carried him over to a group of thirty-five teachers assembled at the front entrance of his mansion.

While Nim's teachers were gathering for their group portrait, I stayed with Nim out of sight in a greenhouse hidden from the house by foliage. The greenhouse was generally off limits to Nim because of the possible damage he could do to its plants and its fixtures. On this occasion, however, I felt that a short visit would do no harm and that it was important to keep Nim away from the largest group of teachers he would ever see until the last possible moment. On signal, I carried Nim out of the greenhouse. Even then I prevented him from seeing his teachers standing in front of the Delafield mansion by holding him below the umbrella that shielded us from the drizzle. When he finally saw his old teachers, Nim's first reaction was to cling to me and hoot softly as if he was both curious and afraid of the crowd he saw. Then he climbed down from my arms and urged me to proceed toward them as quickly as possible.

While pulling on my arm, Nim looked back and forth across the four rows of teachers assembled in front of his home. To insure as formal a picture as possible, I held Nim in my arms for the first few shots, but it seemed unfair to restrain him from being reunited with his old teachers. I also knew that the picture-taking session would be a long one because Laura Petitto, Ronnie Miller, and Steve Hasday, who were delayed in traffic, had yet to arrive. Once I released Nim, he bounded back and forth hugging almost every one of the teachers who had come to say goodby to him. Only two days earlier Nim had been ecstatic about seeing Stephanie and her family for the first time in two years. Now his teachers were sadly bidding farewell to a chimpanzee they all adored.

On the following morning Nim was awakened at 6:00 a.m. Bill and I held him while our consulting pediatrician, Dr. Buddy Stark, injected a sedative (Sernalin and Valium) into his thigh. Five minutes later we wrapped a sleeping Nim in a blanket and were on our way to Teterboro

Some of Nim's old teachers gathered to say goodbye.

Airport in New Jersey, where our plane was to take off at 8:30. Accompanying Nim were Bill, Joyce, Susan Kuklin, and myself. Since space in the plane was limited, Mary Wambach and Jerry Tate (another photographer) had left for Oklahoma the previous day by car.

In the plane, Joyce, Bill, and I took turns sitting next to Nim, who slept on a small shelf in the baggage compartment. My obvious concern during the flight was whether he would wake up and cause trouble. For that reason it was vital that someone sit next to him throughout the flight, to observe when and how forcefully he stirred. I took the first watch and was later relieved by Bill and Joyce.

During my second shift, about two hours after takeoff, Nim sat up and began to whimper. He looked very confused. Shortly afterward he began to scream, apparently frightened by his new surroundings and the disorienting effects of the sedative. Had I not been concerned that Nim's screams would alarm the pilot and the co-pilot, I would have waited longer before readministering the sedative, but I knew that people who were not used to a chimpanzee's screams found them hard to ignore. I therefore decided to sedate Nim again. The second dose of Sernalin was effective until Nashville, where we touched down to refuel. When the aircraft door was opened, the bright sunshine and noise from a nearby plane roused Nim out of his stupor. Again, I had to administer a dose of the sedative. Aside from the two injections en route, the eight-hour flight was uneventful. But that did not mean that Bill, Joyce, Susan, and I were relaxed. Until we touched down in Norman, all of us were fearful that something might go wrong. That, coupled with our sadness over returning Nim, made for a rather grim trip.

When I arrived in Norman, my main concern was how Nim would behave as the heavy dose of the sedative wore off. Various primatologists I had consulted predicted that Nim's motor coordination would be very poor for as long as twelve hours after he woke up. Not knowing what to expect, I had asked Roger Fouts to section off a portion of the building that would be Nim's home at the institute. I would sleep there with Nim on a mattress. I wanted to stay with Nim during his first night in his new home to not only keep an eye on him as the sedative wore off but also because I did not want Nim's first experience of his new home to be alone in a cage, surrounded by other caged chimpanzees.

The only way to approach the building in which Nim and I would sleep was past cages housing other chimpanzees. Accordingly, I kept Nim well out of sight of any of the buildings until it became dark. I then carried him to the area Roger had prepared for me and bedded down with him for the night. We were in the storage section of one of the center's chimpanzee buildings, a space approximately twelve by fifteen feet, separated from the resident chimpanzees by walls made of chicken wire stapled to studs. These walls were covered with black cloth so as to provide visual but not acoustic isolation from the other chimpanzees. Our mattress was placed on the concrete floor in one corner of our area.

After a long day, which for me had begun at 4:30 that morning, and a tiring flight, I was ready to fall asleep. But I soon discovered that I wasn't going to get much sleep that night. Each time it looked as if Nim was about to fall asleep, he began to crawl toward one of the walls next to our bed. I tried to restrain him but was only moderately successful. After Nim pushed against the wall momentarily, he relaxed and I was able to pull him back toward me. Again I tried to fall asleep with one arm holding Nim, and again he started crawling, this time toward the wall at the head of our bed. Yet another time I pulled him back, but not before he literally bounced off the wall. These cycles repeated themselves at what seemed to be fifteen-minute intervals. Nim's movements thrashing around the bed did not seem directed. If anything, they were somewhat spasmodic, as if his body was trying to shake off the effects of the sedative. As far as I could tell, Nim was asleep during each of these crawling episodes. He continued to move around in our bed until at least 3:30 in the morning. That was the last time I recall looking at my watch.

At about 7:30 I tried to rouse Nim. I wanted to get him out of the building before 8:00 a.m., when the caretakers arrived to feed the other chimpanzees of the colony. I had been warned that pandemonium broke out when the chimpanzees saw the feeders' car drive up each morning. Since Nim had yet to encounter another chimpanzee, I thought I would spare him the ear-splitting cries of more than forty of them whooping it up in anticipation of breakfast. Even though Nim kept pulling me back toward our bed, I managed to get him out of the building just before the feeders arrived. I carried him for about a quarter of a mile, until we reached the wooded area where I sat with him the previous night.

Nim's first glimpse of another chimpanzee actually occurred then, as I carried him from the building. One of the cages we had to pass housed Washoe, the first chimpanzee to learn some rudiments of sign language, and three other females. It was impossible for Nim not to notice Washoe and the other three chimps as we walked by. He looked in their direction momentarily but showed no reaction. He then looked away as if he had seen nothing strange. Having known Nim for almost four years, I wasn't fooled. Many times I had seen him pretend not to notice something that absolutely fascinated him. Later, at a time that apparently seemed safe to him, he would direct his full attention to the object he had pretended to ignore.

That time arrived soon after I put Nim down in the wooded area. I had expected him to curl up in my lap and go back to sleep. Instead, he took my hand and began to lead me through a grove of apple trees situated in front of Washoe's cage. Under the cover of a tree, he stared directly at Washoe, grunted softly, and pointed in her direction. His gaze was intent. Nim held on to the tree with one hand and to my hand with the other, staring and grunting for about five minutes. He was clearly intrigued by the sight of Washoe. At the same time he needed the security provided by the cover of a tree and my presence. Abruptly, he pulled my hand and led me to the next tree in the grove, bringing us about ten yards closer to Washoe, who could not see Nim and who would probably have been indifferent if she had. This surveillance of Washoe continued for another half hour. By that time Nim was under the tree closest to Washoe's cage. When he seemed relaxed, I tried to lead him away from the tree, closer to Washoe, but he strenuously resisted each attempt. He clearly preferred to satisfy his curiosity about Washoe from a distance.

Joyce and Bill arrived at 9:00 and shared breakfast with Nim in a grassy area about thirty yards from Washoe's cage. With Nim shaking off the effect of the sedative in the watchful company of Joyce and Bill, I was free to drive back to Norman to shower and get some breakfast. Later that morning I met with Roger Fouts to plan Nim's first face-to-face encounter with another chimpanzee. Our original plan was to have Nim meet Ally, his eight-year-old brother, whom Roger had taught to sign. Our plan was to see whether, once Nim got over the shock of meeting another chimpanzee, he and Ally would sign to each other. All of this was supposed to be filmed by a television crew making a short documentary about language in chimpanzees. Roger and I and Harry Moses, the producer of the program, had previously agreed not to film the very first meeting of Nim and Ally. Even though it would be of historical interest, we felt that it might be best to wait a day in order to give Nim enough time to feel comfortable with another member of his species.

After seeing how Nim avoided Washoe, even when she was behind the bars of a cage, it seemed doubtful that Nim would be able to handle a face-to-face encounter with his own brother, a chimpanzee with a reputation both for strength and for provoking fights. Roger suggested a younger and more mild-mannered chimpanzee, a seven-year-old male named Mack. Mack was quite small for his age and was considered to be one of the gentlest chimps of the colony. While Mack was not as proficient a signer as Ally, he did know about twenty signs, many of which coincided with Nim's vocabulary. Roger and I planned to try a low-key meeting of Nim and Mack sometime after lunch. I would sit with Nim in an open grassy area, and Roger would walk over with Mack. Both chimps would be on long leads that would allow them to move around freely but be restrained if necessary.

The first chimp Nim was introduced to: Mack

Despite my attempts to relax Nim by tickling and chasing him, his body tensed as soon as he saw Roger and Mack approach from a distance. He grunted and positioned himself at my side with one arm around my neck, presumably in readiness to hide behind my back or run away if Mack approached too closely; in any case, that is precisely what he did when Mack and Roger sat down about ten feet away from us. First Nim sat down behind me and peered at Mack while clutching at my chest. Then he ran in the opposite direction, pulling tightly on the lead. I pulled him back and had my hands full trying to hold him next to me. Sensing that I wasn't going to let him run away, Nim signed *dirty* repeatedly, as if he needed to go to the toilet. The mendacity of that sign was transparent. Nim was simply trying everything in his power to remove himself from an unpleasantly strange situation. After a few minutes Nim settled down in my lap and looked defiantly toward Roger and Mack.

Roger and I decided to wait and see what happened without forcing anything between the two chimpanzees. Mack looked at Nim, but he did not at first try to approach him. When he did, about fifteen minutes later, Nim sprang into a defensive posture and struck out at Mack in a way that reminded me of the way he attacked bugs.

Nim's first meeting with Mack. Mack crouches in front of Joyce Butler, and the author holds Nim's lead.

Roger, I, and the chimpanzees continued to sit across from one another for about a half hour. During that time we made some other futile attempts to introduce Nim to Mack. I tried tickling Nim, and while he was flat on his back, I beckoned to Roger to approach with Mack. At the last moment, however, Nim sprang up and pushed Mack away from him. I also tried to feed Mack a few times, an act I was sure that Nim would resent. I hoped that Nim would get close enough to Mack to stare at him eating or to try to get some of the candy I had given him. I had often witnessed Nim react that way in New York when his caretakers shared food but excluded him. But his first face-to-face encounter with another chimpanzee was clearly not that kind of situation.

After exhausting all our ploys to have Nim accept Mack, Roger and I agreed to try again that same afternoon. Our second effort turned out to be almost a carbon copy of the first. Nim seemed to fear Mack and presumably chimpanzees in general. I told Harry Moses and his crew, who had arrived just after Nim's second encounter with Mack, that it would probably take a few weeks before our city boy would settle down in his farmlike surroundings and overcome his fear of other chimpanzees. Harry appreciated the problem but felt we should proceed the next morning as planned.

My main worry at the end of Nim's first day in Oklahoma was not his reaction to Mack but where he would sleep that night. Up to this time Nim had always slept with a caretaker or in a familiar bedroom. From now on he would have to sleep in a cage and stay in the cage for many hours each day. The cage reserved for Nim was the most spacious of the four in his building and was situated across from a row of cages housing three other chimpanzees: Lilly, a new female who was also raised in a home, and Kelly and Mack, two males born at the center.

I tried to prepare Nim for his new life by carrying him to his cage. Almost as soon as I brought him into the building he struggled to get out of my arms. When I put him down he scooted as far away from me as his lead would allow, about fifteen feet outside the building. Trying to pull him back into the building would have been a difficult tug of war that I probably could have won, but it would have been foolish to do so. Since Nim would be living in this building, there was no point in forcing him to enter it unless all other efforts failed.

Second meeting. Nim is no more eager than before.

I decided to see if Nim would take the initiative and try to explore this strange building on his own. But each time I gave Nim an opportunity to approach it, he ran away as far as he could. It soon became clear that Nim was not going to go out of his way to enter his new quarters. I decided that the next best thing was to lure him inside. After his second fruitless session with Mack, I asked Joyce, Bill, and Mary to sit with me inside the building just outside Nim's cage. Nim stayed outside the building and signed to us *come*, *play*, and *there*, pointing away from the building. We all did our best to lure him inside by playing with some of his favorite playthings, but he remained outside, staring at the ground plaintively. I then suggested that we turn our backs to him and start playing with Lilly, the chimpanzee in the cage closest to Nim's. That strategy paid off rather quickly. Nim slowly walked in and sat down next to us.

There still remained the task of getting Nim to enter his cage, which proved easier than I expected. I asked Bill to go inside the cage and act as if he was having a good time. Bill played the role perfectly. He swung from the ceiling of the cage, ran around it a few times, and banged excitedly on the bars. Nim rushed into the cage partly out of curiosity and partly because entering the cage with Bill must have seemed a good way of increasing his distance from the other chimpanzees across the way. Bill and Nim played around inside for about five minutes. In the process Bill slipped off Nim's harness and lead. That precaution was taken out of concern that Nim might choke himself by getting tangled in the lead and then pulling on it too tightly in his struggle to get free.

While Nim was frolicking around in his cage, I motioned to Bill to make a quiet exit at the first opportunity. A few minutes later Bill slipped out the door. Since I had entered the building I had not moved from my original position just outside the door of Nim's cage. Having experienced many unsuccessful attempts to remove myself from Nim abruptly, I knew that I would have only one chance to do so on this occasion. Once Nim caught on that he was to be left by himself it would be virtually impossible to lock him in his cage.

Many times I had prepared myself to say goodby to Nim, but no amount of preparation could have made the next few minutes less painful. As Bill walked out of the cage, I slowly swung the door shut. Nim didn't realize what had happened until I got up and padlocked the door. He then began to scream and tried to force the door open. I asked everybody to leave immediately. If there was to be a change in Nim's life, there was no point in adding to his frustration by prolonging our presence. Without further ceremony we all walked out of the building.

I will never forget Nim's incessant, ear-piercing screams and his look of fear and anger when I abandoned him in his cage. This was by no means the first time I had heard Nim scream, nor was it the first time I had seen him outraged at the cruel realities of his socialization. Nim often threw a tantrum when he was prevented from being with a favorite teacher, when he was not included in some activity, or when he was punished for bad behavior. In each of these situations there was always someone around to comfort him and to try to get him to understand why his caretaker acted as he or she did. But on this occasion there was no way for Nim to understand why he had abruptly been left by himself in a totally strange environment.

Even though I knew I would see Nim the next day, there was no question that my exit that evening was my farewell. I felt a great loss. Despite the frustrations of Project Nim, I knew that there could be no substitute for that intelligent bundle of playfulness and mischief, a creature more human than any other nonhuman I knew. One of the reasons this parting was so painful was that there was no way to talk with him about it. Nim and I were able to sign about simple events in his world and mine. But how could I explain why I and the other project members who came along to Oklahoma suddenly abandoned him? How could we explain that it was necessary to leave him forever in a totally new environment, with a totally new group of human and nonhuman primates?

That evening, while having dinner with Harry Moses and the film crew, I received a call from Dr. Lemmon. Before turning in for the night, Dr. Lemmon had gone to Nim's building to see how he was doing. To his surprise, Nim had broken the pipe above his cage, which Dr. Lemmon had assumed was chimp-proof. The problem was remedied for the moment by shutting a valve and allowing the pool of water that had collected in Nim's cage to flow into a drainpipe in the floor. Dr. Lemmon, who had presided over more than a dozen rehabilitations of chimpanzees returned to the institute after they had been reared in human homes, told me that he had never witnessed so violent a reaction to a separation between a chimpanzee and its surrogate family. Even before I had arrived with Nim in Oklahoma, Dr. Lemmon had tried to assure me that our separation would proceed without a hitch. He was clearly surprised at Nim's reaction to his new existence. I wasn't.

Before daybreak the next morning, Joyce, Bill, Mary, Susan Kuklin, Jerry Tate, Harry Moses, his crew, and I left our motel rooms and drove out to the center. When we arrived the full moon in the west was just setting, almost precisely opposite the point on the horizon where the sun was rising. Bill and Joyce got Nim up just after the sun rose. When they entered his building he was sound asleep on a shelf hung from one of the walls of his cage. Nim showed no trace of his violent tantrum of the prior evening, and he greeted his teachers affectionately and ate a hearty breakfast.

The site of Nim's third introduction to Mack was a remote field about a quarter of a mile south of the main compound of the primate center. After his breakfast I led Nim to that field and played with him while the television cameramen set up their equipment.

This time Nim seemed less afraid of Mack when he and Roger sat down next to us. But just as Nim had done the previous day, he avoided Mack each time he came too close. I was not very optimistic that the outcome of this meeting would be different from that of Nim's first two meetings with Mack. Several times, I tickled Nim in an effort to relax and distract him, but he must have sensed what I had in mind. Each time Mack got too close, Nim would pull back and shoo Mack away. Once, Mack was able to poke Nim before he was able to prepare himself. Nim jumped back violently but otherwise did not react to this first instance of being touched by another chimpanzee.

I gave up on tickling for the moment and tried some other approaches. I borrowed Roger's pipe, smoked it for a while, and gave it to Mack. Nim watched intently, looking first at me and then at Mack. He obviously wanted to smoke the pipe himself, but all he could do was stare at us with a look of hurt and wonder, as he had the previous day when I had offered candy to Mack. Nim looked as if he could not believe that I had given candy or a pipe to another chimpanzee, and he reacted in almost the same way when I gave Mack a baseball cap to wear.

Shortly after the cameraman changed film and shifted his position, I again tickled Nim and moved him quite close to Mack, who at that moment was being tickled by Roger. Not knowing what to expect, I moved Nim so that his back brushed up against Mack's. That touch seemed to reverse magically all the failures of the previous day and that morning. Abruptly, Nim and Mack began scuffling furiously. I couldn't have been more delighted. Nim and Mack both had big grins on their faces and were obviously having a wonderful time. From that moment on, it was hard to keep Nim and Mack apart. In a way that only chimpanzees can, Nim and Mack played with each other insatiably. They tickled each other, rolled over a few times, and then resumed flailing their arms at each other.

Once Nim and Mack took to each other I felt that the main goal of this encounter had been realized and I could return to New York confident that Nim had at least one playmate in the Oklahoma colony. Yet to be resolved, however, was whether Nim and Mack would try to communicate with each other in sign language. At that time I wasn't very optimistic about it, but I had worked out with Roger a few simple situations in which I thought Nim would be motivated to sign. I asked Roger to pull Mack toward him, and I did the same with Nim. Roger signed *sign* to Mack, to which Mack responded by signing *come*. I signed *hug?* to Nim,

Above: third meeting. Friends at last. Below: Roger: *sign* / Nim: *hug*

and he responded by signing *hug* repeatedly. On at least two occasions Nim signed *hug* without any encouragement from me. I was also able to get Nim to sign *hat* by placing a baseball cap on Mack's head and restraining Nim. Without any prompting or signing on my part, Nim signed *hat*. When I released Nim, he jumped at Mack and began an unsucessful attempt to wrest the hat away from him.

Nim signs *hat* when there is nothing else he can do (top) . . . but obviously prefers action to words.

Everyone was thrilled at what seemed to be two clear instances of one chimpanzee signing to another. While I was encouraged by what I saw, I remained somewhat dubious. I felt that the real test of communication between the two chimpanzees had to be made without any humans present. While watching Nim sign *hug* and *hat* I could not be sure whether he was signing to me or to Mack. Given Nim's long history of trying to satisfy me and other teachers by signing, I could not rule out the possibility that his signs were directed at me.

After the television filming was completed we packed our gear in readiness to return to New York. Nim was left with Roger Fouts and one of the caretakers from the center just outside the building housing his cage. Nim remained calm so long as we were in view. While we loaded our cameras and other belongings into our cars, Nim made a few attempts to approach us. Halfheartedly we responded to some of these overtures.

But neither I nor any of his teachers felt that this was a moment for fun and games with Nim. What bothered me was not simply that I was leaving someone whom I had grown to love during the four years I had known him so well. Nor was it simply a feeling that Nim deserved something better than the environment in which I was leaving him. At this well-managed primate center I was sure that Nim would be looked after as well as any of the other resident chimpanzees. What concerned me most was the fact that Nim would lack the opportunity to sign and to be signed to sixteen hours a day, seven days a week, and that he would be deprived of the human companionship that he had come to expect during all of his waking hours. I was aware that Roger Fouts was interested in doing research on sign language in chimpanzees and that he and his graduate students would occasionally try to sign with Nim. Bill and Joyce had made sure that all of them understood the extent of Nim's knowledge of sign language. But I could not be very optimistic about the stimulation that Nim would receive. The contact that Roger and his students would have with Nim would be limited to an hour or two a week.

In short, I had to acknowledge that I was abandoning a wonderful little creature of high intelligence to an institutional existence. Whatever potential Nim had to develop his intelligence and to learn more language would almost certainly never be fulfilled. It was with real sadness that we drove away from the center. Nim was with Roger outside his cage when we left. He did not appear to be particularly upset. It would probably take him a while to understand how profoundly his life had changed.

13

Some Unexpected Discoveries

Project Nim continued for another year after Nim was returned to Oklahoma but in a much subdued manner. The all-out and often chaotic effort needed to watch Nim and to teach him to sign was replaced by the more leisurely pace of a typical scientific project. No longer was there an urgent pressure to recruit and train new teachers or to make sure that there was always someone available to be with Nim. I could now come home at night without the dread of finding a message that a teacher was sick or unavailable for his or her session the next day because of some other emergency. Best of all was the time I had to analyze the volumes of data that had accumulated during the last few years, to contemplate what Nim had actually learned about sign language, and to write about our findings.

As much as I valued my newly acquired freedom, I felt sad that its price was the abandonment of Nim to a life quite different from the one I had provided for him in New York. From the start of the project I had known that it was highly unlikely that I would be able to keep Nim as long as he lived. But it was not possible to predict the circumstances under which I would have to give Nim up. Intellectually, I could tell myself that there was little point in trying to keep Nim without the resources I needed to advance the scientific output of the project. But that thought could not offset the sense of loss I felt over not having Nim around any longer, over leaving him in an unstimulating environment, and over not being able to push the project as far as I had originally hoped.

Through letters and phone calls, I learned that Nim's adjustment to the primate center was slow but steady. After what must have been a very trying quarantine period, in total isolation from other chimpanzees, Nim was placed in a large cage housing Mack and two other juvenile chimpanzees. Occasionally he was taken for walks by the caretakers of the center. As far as I could tell, little effort was made to sign with Nim, but the biggest change for Nim was the fact that he was now caged and in the company of chimpanzees rather than humans. It would have been intriguing to be able to ask Nim which he preferred.

With Nim no longer around, I experienced an eerie feeling when I went back to Delafield or the classroom. When Nim was living at Delafield, he would suddenly bolt into view or race toward me when I arrived. Now, it seemed as if the energy level of the beautiful estate had dropped abruptly. In the classroom I felt the same way. With Nim around, I had paid little attention to the cold cinder block walls of the complex. Without Nim, I wondered how I and the other teachers could have spent so much time in these oppressive rooms.

Within a month of Nim's move to Oklahoma, Dick, Bill, Joyce, and I had begun the lengthy task of sifting through more than 2,000 teachers' reports in order to extract whatever information we could about Nim's day-to-day usage of sign language. We also began to transcribe the more than forty videotapes that had been made of Nim's conversations with his teachers. On the average, it took an hour to transcribe one minute of tape. Since we had more than forty hours of tape and each transcript had to be checked by at least two teachers, the job would take at least a year.

Within three months of Nim's return to Oklahoma, I had seen enough interesting material in our data to console myself that Project Nim had advanced our understanding about what a chimpanzee can learn about sign language. This was true both in the case of individual signs and combinations of signs. Nim's vocabulary included signs that seemed to serve as substitutes for aggressive behavior (*bite* and *angry*). Another significant and heretofore unreported use of sign language was the manipulation of the behavior of his teachers, as with *dirty* and *sleep*. As far as I knew, this was the first observation of an ape's using an arbitrary symbol to misrepresent a bodily state. It might be tempting to argue that Nim was lying when he signed *dirty* when he did not have to go to the bathroom or *sleep* when he wasn't sleepy, but that interpretation seems too strong because there is no evidence that Nim was aware of the difference between a true and a false utterance. Nim did learn, however, with no help from his teachers, that signs were powerful tools and that he could manipulate the behavior of his teachers by misrepresenting the meaning of certain signs. Project Nim also showed that a chimpanzee's acquisition of new signs went through an orderly sequence of stages, some of which resembled the stages a child goes through as it masters a new word. Nim's errors in using signs often appeared to result from confusion, not about the meaning of a sign, but about its form: signs that were topographically similar to other signs were often used instead of signs that were similar in meaning. A similar phenomenon has been observed in both spoken language and ASL. Finally, Nim often signed spontaneously, without food or drink rewards, about pictures in order to identify what he saw. He was often observed to sign to himself, for whatever intrinsic pleasure that produced, while flipping through a book or magazine with his back to his teacher.

The main goal of Project Nim was to provide information about a chimpanzee's ability to create a sentence. As far as I know, our corpus of Nim's combinations is unique in studies of sign language in apes. It is without parallel in studies of language learning by either deaf or speaking children. The regularities in our corpus that were noted before Nim was returned to Oklahoma gave me reason to believe that Nim was creating primitive sentences. Our intensive post-Oklahoma effort at data analysis had hardly begun, however, when I began to doubt that Nim's combinations were legitimate sentences.

One of the first facts that troubled me was the absence of any increase in the length of Nim's utterances. During the last two years in which Nim lived in New York, the average length of his utterances fluctuated between 1.1 and 1.6 signs. That state of affairs characterizes the *beginning* of a child's development in combining words. As a child gets older, the average length of its utterances increases steadily. This is true both of children with normal hearing and of deaf children who sign. Having learned to make utterances relating a subject and a verb (such as "Daddy eats") and utterances relating a verb and an object (such as "eats breakfast") the child apparently learns to link them into longer utterances relating the subject, verb, and object (such as "Daddy eats breakfast"). Later, the child learns to elaborate that utterance into statements such as "Daddy didn't eat breakfast" or "When will Daddy eat breakfast?" and goes on to still further elaborations. Despite the steady increase in the size of Nim's vocabulary, the mean length of his utterances did not increase. Apparently, utterances whose average length was 1.5 signs were long enough to express the meanings that Nim wanted to communicate.

Even though the mean length of Nim's utterances did not exceed 1.6, some of his utterances were very long. Yet they were as a rule very repetitious. Consider, for example, the longest utterance Nim was observed to make, an utterance containing 16 signs: *give orange me give eat orange me eat orange give me eat orange give me you.* Whereas a child's longer utterances expand the meaning of shorter utterances, this one does not. Further, the maximum length of a child's utterance is related very reliably to its average length. Nim's longer utterances neither added new information, nor was the maximum length of his utterances related to their average length.

Our corpus of Nim's combinations allowed us to observe in detail the nature of his progression from two- to three-sign combinations. Table 10 shows the twenty-five most frequent two- and three-sign combinations, and their absolute frequencies. From a lexical point of view, a comparison of these combinations reveals that the topics of Nim's three-sign combinations overlapped considerably with the topics of his two-sign combinations.

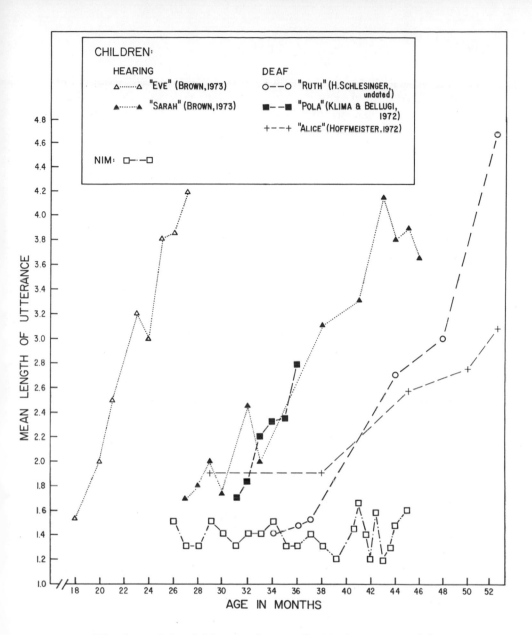

Unlike those of the children in this sample, Nim's utterances did not increase significantly in length as he grew older.

His three-sign combinations do not, however, provide new information. Consider, for example, Nim's most frequent two- and three-sign combinations: *play me* and *play me Nim*. Adding *Nim* to *play me* is simply redundant. A further complication is revealed when one realizes that the three-sign combination *play me Nim* may have been derived by adding the single sign, *play*, to Nim's second most frequent two-sign combination, *me Nim*. There is no obvious way to choose between the two derivations of *play me Nim* suggested by Table 10: *play me* + *Nim* and *play* + *me Nim*. Similar alternatives present themselves when trying to derive the other three-sign combinations shown in Table 10.

Table 10.
Most Frequent Two- and Three-Sign Combinations

Two-Sign Combinations	Frequency	Three-Sign Combinations	Frequency
play me	375	play me Nim	81
me Nim	328	eat me Nim	48
tickle me	316	eat Nim eat	46
eat Nim	302	tickle me Nim	44
more eat	287	grape eat Nim	37
me eat	237	banana Nim eat	33
Nim eat	209	Nim me eat	27
finish hug	187	banana eat Nim	26
drink Nim	143	eat me eat	22
more tickle	136	me Nim eat	21
sorry hug	123	hug me Nim	20
tickle Nim	107	yogurt Nim eat	20
hug Nim	106	me more eat	19
more drink	99	more eat Nim	19
eat drink	98	finish hug Nim	18
banana me	97	banana me eat	17
Nim me	89	Nim eat Nim	17
sweet Nim	85	tickle me tickle	17
me play	81	apple me eat	15
gum eat	79	eat Nim me	15
tea drink	77	give me eat	15
grape eat	74	nut Nim nut	15
hug me	74	drink me Nim	14
banana Nim	73	hug Nim hug	14
in pants	70	play me play	14
		sweet Nim sweet	14

The overlap between Nim's most frequent two- and three-sign combinations is apparent in other comparisons as well. Eighteen of Nim's twenty-five most frequent two-sign combinations can be seen in his twenty-five most frequent three-sign combinations, in virtually the same order in which they appear in his two-sign combinations. A striking similarity emerges between Nim's two- and three-sign combinations if one considers only the signs and not their order of occurrence. All but five of the signs that appear in Nim's twenty-five most frequent two-sign combinations appear in his twenty-five most frequent three-sign combinations. The five exceptions are *gum, tea, sorry, in,* and *pants.* The combination *in pants* was the least frequent two-sign combination shown in Table 10. It occurred mainly during dressing and after trips to the toilet.

The repetition of the same sign, as for example in *eat Nim eat* and *nut Nim nut,* is another redundant feature of Nim's three-sign combinations. In producing a three-sign combination, it appears that Nim was simply adding emphasis. Nim's four-sign combinations reveal a similar picture. Table 11 shows all four-sign combinations whose frequency is equal to or greater than three. Fifteen of the twenty-one types of four-sign combinations shown in Table 11 contain repetitions of some signs, for example, *eat banana Nim eat* and *grape eat Nim eat.* If *me* and *Nim* are equated, on the grounds that they mean the same thing, twenty of the twenty-one combination types shown in Table 11 repeat the same sign. That leaves but one combination type, *me eat drink more,* which contains four distinctly different signs. Seven of the twenty-one combinations shown in Table 11 repeat two-sign combinations in the same order, for example, *drink Nim drink Nim* and *me gum me gum.*

Table 11.
Most Frequent Four-Sign Combinations

Four-Sign Combinations	Frequency	Four-Sign Combinations	Frequency
eat drink eat drink	15	drink eat me Nim	3
eat Nim eat Nim	7	eat grape eat Nim	3
banana Nim banana Nim	5	eat me Nim drink	3
drink Nim drink Nim	5	grape eat me Nim	3
banana eat me Nim	4	me eat drink more	3
banana me eat banana	4	me eat me eat	3
banana me Nim me	4	me gum me gum	3
grape eat Nim eat	4	me Nim eat me	3
Nim eat Nim eat	4	Nim me Nim me	3
play me Nim play	4	tickle me Nim play	3
drink eat drink eat	3		

The lack of growth of the length of Nim's utterances and the numerous repetitions of signs in his longer utterances distinguish Nim's use of language from that of a child. These lexical features of Nim's signing suggest a considerably weaker view of his linguistic competence than does the evidence I noted in Chapter 11. In reviewing our semantic analysis I became skeptical about its validity for two reasons. One problem is the small number of signs used to express particular semantic roles. Ninety percent of the combinations interpreted as an expression of location involved the sign *point*. A similar state of affairs exists in the case of combinations interpreted as expressions of recurrence. That role was represented exclusively by *more*. In combinations presumed to relate an agent and an object or an object and a beneficiary, one would expect a broad range of examples—*Nim, me, you*, and names of other animate beings. However, 99 percent of the beneficiaries in utterances judged to be object-beneficiary combinations were *Nim* and *me*, and 76 percent of the agents in utterances judged to be agent-object combinations were *you*. There were other examples showing the limited range of signs used by Nim to express different semantic categories. This is in contrast to the results of semantic analyses of children's utterances that normally reveal a greater richness of words within different semantic categories, and an orderly progress toward greater complexity. The issue of progress is irrelevant in the case of Nim's combinations because the number of his utterances containing three or more signs was too small to warrant a serious semantic analysis.

The other reason for doubting the conclusion that Nim's two-sign combinations were simple sentences came from a detailed analysis of videotapes. These tapes provided me with an invaluable opportunity to analyze, with the wisdom of hindsight, just about any aspect of Nim's signing. The importance of this becomes clear when one recalls that the data upon which the distributional analyses were based came solely from teachers' reports. Whatever the teacher missed about Nim's signing was missed forever. The videotapes, on the other hand, captured it all.

It was while looking at playbacks of videotapes that I realized I had missed an important aspect of the context of Nim's signs. When the other teachers and I were working with Nim and recording what he signed, our attention had always been riveted on his signing. We had paid too little attention to what *we* signed to Nim. It is true that teachers were attentive to what they signed in determining whether Nim satisfied our criterion for learning a sign, a criterion that required that Nim make a particular sign without the teacher having made it during his or her prior utterance. But we did not ask systematically how often Nim's utterances contained signs that we had just signed to him. We also did not consider how often Nim signed if his teacher hadn't first signed to him. I was aware that Nim often signed spontaneously to request and describe things, but until we began to glean that information from our videotapes we could only guess how frequently Nim signed spontaneously.

Studies of language development in children, referred to as discourse analyses, have revealed a number of systematic relationships between a child's utterances and those of its parents. Initially, it is often the parent who starts the conversation. One study estimated that 70 percent of a child's utterances were occasioned by something the parent had said. In most instances the child did not reply by simply repeating what the parent had said but added to the parent's utterance or created a new one from totally different words. Less than 20 percent of a child's utterances were *imitations* of its parent's utterance. The remainder of the child's replies were either elaborations of the parent's utterance or totally novel but relevant utterances.

Our initial analysis of the relationship between Nim's utterances and those of his teachers showed that Nim's were more dependent upon what his teachers signed than a child's are dependent on the words of its parents. In order to characterize the relationship between Nim's utterances and those of a child, we assigned each of Nim's utterances to one of five categories: spontaneous, imitation, reduction, expansion, or novel. A spontaneous utterance was one that was not preceded by a teacher's utterance. All of the remaining categories included utterances that were preceded by a teacher's utterance. Imitations were exact reproductions of his teacher's signs. Reductions were imitative utterances that omitted some of the teacher's signs. An expansion included some of the teacher's signs and some new signs added by Nim. A novel utterance was one in which there was no overlap between the signs of a teacher's utterance and Nim's response.

Some photographs of videotapes show how Nim's signing was related to that of his teachers. Below and on the next page we see Nim making both spontaneous and novel signs. The dialogue between Nim and Joyce

Nim: *eat*

Nim: *sweet*

Nim: *red*

Nim: *red*

Joyce: *sweet?* / Nim: *thirsty*

Joyce: *sweet?* / Nim: *thirsty*

was about candy. Nim wanted Joyce to give him some red candy, hence his spontaneous sequence: *eat sweet red*. Joyce was trying to clarify what Nim signed by checking to see if he actually wanted a sweet. Nim responded by continuing the sign for *red* (the index finger touching the lower lip) and then touching his throat as if to sign *thirsty*. Joyce concluded that Nim's *thirsty* sign was a topographical error. The difference between *thirsty* and *sweet* has mainly to do with where the fingers touch the body. In the case of *sweet*, the index and middle fingers touch the lower lip. *Thirsty* is signed by moving the index finger downward while touching the throat.

On the next page, Joyce is telling Nim to choose the red can. Nim signs *red* and then chooses the red can. In this example, Nim's sign was imitative of Joyce's.

Next we see an example of partial overlap between Nim's signing and the teacher's signing. Dick had previously given Nim a drink from the red cup and wanted Nim to sign about its color, hence his statement *that red*. Nim was requesting another drink. His request included both of Dick's signs, but in reverse order and with a new intervening sign, an example of an expansion.

Joyce: *choose*

Joyce: *red*

Joyce: *red* / Nim: *red*

Nim chooses the red can.

Dick: *that*

Dick: *red*

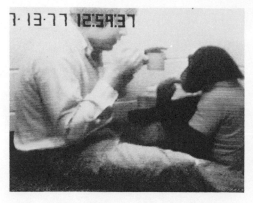

Nim: *red*

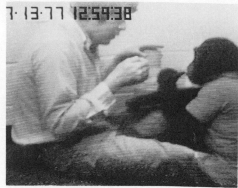

Nim: *drink*

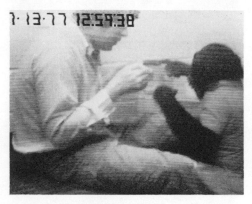

Nim: *that*

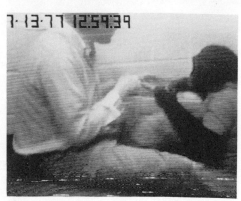

Nim: *more*

During Nim's last year in New York only 10 percent of his videotaped utterances were spontaneous. Approximately 40 percent were imitations or reductions. If the conversations we videotaped and transcribed were representative of the thousands of conversations from which our corpus was derived—and I have no reason to believe that they were not—I must conclude that Nim's utterances were less spontaneous and less original than those of a child. To a much larger extent than a child's, Nim's utterances were variations of the signs contained in his teacher's prior utterance. He was much less likely than a child to add new information to a conversation in his replies. (I should note that it would be more valid to compare Nim with a deaf child than with a speaking child. But until more data become available showing how deaf children learn to sign, the only way to compare Nim's use of language with that of a child's is to use data obtained from speaking children. The limited data that is available show no major differences in language acquisition by hearing children who speak and deaf children who sign.)

An unanticipated, but instructive, example of the influence of the teacher's signing on Nim's signing can be seen in the photographs on page 173, where Nim is shown signing *me hug cat.* A careful examination of these photographs reveals that Nim's teacher, Susan Quinby, signed *you* while Nim was signing *me,* and *who?* while Nim was signing *cat.* Because these were the only four photographs taken of this conversation, we cannot be certain just when Susan began her signs. It is not clear, for example, whether Susan signed *you* simultaneously or immediately prior to Nim's *me.* It is, however, unlikely that Susan signed *who?* after Nim signed *cat.* Inspection of other photographs revealed a similar pattern of conversation between Nim and his teachers. For example, in the picture at top right on page 148, where Nim is signing *black,* Susan is also signing; she is asking the question, *What's that?* On page 149 Nim is shown signing *Nim* in response to Susan's query *who?* On page 121 Nim is shown signing *hug* in response to Jean's prompting use of that sign. On page 177 Nim is signing *me* in response to my signing *mine.* These photographs underscore the importance of a discourse analysis for revealing the extent to which Nim's utterances were influenced by his teachers' signing.

Aside from Nim's tendency to rely on his teachers' signs as a basis for his own signing, our analysis of videotapes and other data revealed two other important differences between Nim's and a child's use of language. One has to do with the ability to take turns in a conversation. During a conversation with a parent or other adult, children show a good sense of when to listen and when to talk. This has been demonstrated by the relatively low frequency with which a child interrupts a speaker's utterances. Our videotapes presented a different pattern in Nim's exchanges with his teachers, a pattern that previously was not apparent to his teachers or to observers. By playing the tapes in slow motion it was possible to see just when Nim began an utterance in relation to his teacher's utterance. Our analysis showed that Nim interrupted his teachers much more frequently than a child interrupts its parents. After seeing the extent to which Nim interrupted his teachers, it appeared that he was more concerned about telling his teachers what he wanted of them than he was about what his teachers were saying to him or about exchanging information with them.

I have strong reason to believe that Nim's signing with his teachers is not exceptional in this respect. There is evidence in two films about the Gardners' work with Washoe supporting the idea that prompting has played a much greater role than heretofore recognized in conversations between chimpanzees and humans. One film, a forty-five-minute documentary produced by "Nova," is entitled *The First Signs of Washoe;* the other, produced by the Gardners themselves, is *Teaching Sign Language to the Chimpanzee: Washoe.* Some of the same conversations are shown in both films, but at greater length in the latter. An examination of the conversations in the two films illustrates vividly the need for recording the complete context of a human-ape discourse.

In one example, the "Nova" film shows Washoe signing with Beatrice Gardner. Mrs. Gardner signs *what time now?* and Washoe interrupts to sign *time eat, time eat.* A longer version of the same exchange, shown in the other film, makes it plain that the conversation began with Mrs. Gardner signing *eat me, more me,* after which Washoe gave her something to eat. Then she signed *thank you*—and only then asked *what time now?* Under the circumstances, Washoe's response *time eat, time eat* can hardly be considered spontaneous, since Mrs. Gardner had just used the same two signs, and because it was in answer to her direct question.

The potential for confusion in inadequate reporting is made plain by another exchange that appears in both films. Washoe is with her teacher Susan Nichols, who has a cup and a doll. Ms. Nichols points to a cup and signs *that.* Washoe signs *baby.* Ms. Nichols brings the cup and the doll closer to Washoe, allowing her to touch them, then slowly pulls them away, signing *that* and pointing to the cup. Washoe signs *in* and looks away. Ms. Nichols brings the cup and doll closer to Washoe again, who looks at the two objects once more and signs *baby.* Then, as Ms. Nichols brings the cup still closer, Washoe signs *in. That,* signs Ms. Nichols, and points to the cup. *My drink,* signs Washoe. Now the question is, is this utterance by Washoe—*baby in baby in my drink*—either spontaneous or a significant, creative use of words? It is actually a "run-on" sequence with very little relationship among its parts. Only the last two signs were uttered without prompting on the part of the teacher. The sequence of the prompts, moreover (pointing to the doll and then pointing to the cup), follows the order called for to construct an English prepositional phrase. In short, careful analysis makes the chimpanzee's linguistic achievement less remarkable than it might first seem.

There is no reason to believe that these films, limited as they are, show Washoe at much less than her best. One could wish for more comprehensive records (as far as I know, these are the only films of apes signing publicly available), but nothing in them suggests anything other than a consistent tendency for the teacher to initiate the signing and for the signing of the ape to interrupt and mirror the teacher. *Teaching Sign Language,* the longer of the two films, contains 155 of Washoe's utterances. One hundred and twenty were single-sign utterances, and occurred mainly in vocabulary testing sessions. Every one of Washoe's multisign sequences (24 two-sign, 6 three-sign, and 5 four-sign sequences) was preceded by a similar utterance from her teacher. The "Nova" film also included short segments about Nim's brother Ally and the gorilla Koko.

It shows that all of Koko's and most of Ally's utterances (in each case, single signs) were prompted.

In this light, the distributional regularities observed in the corpus of Nim's two-sign combinations cannot be considered solid evidence of Nim's ability to combine two or more words to create specific meanings. Even though I could rule out rote learning and direct imitation as explanations of the regularities I observed, I could not rule out strategies that would create sequences that looked like sentences but that could be explained by simpler processes. Recall, for example, Nim's preference for *give* + X over X + *give*. Maybe Nim was simply signing *give* and one of the signs the teacher signed to him. Since it was not possible to record both Nim's and the teacher's signs in the reports that served as a basis for our corpus, there is no direct way of showing just what strategy Nim followed in producing this and other regularities the corpus contained. However, I remember asking Nim questions such as *Nim want book?* or *Nim eat apple?* in order to initiate conversations with him, and so does every other teacher with whom I have spoken. In order to sign *give* + X, all Nim had to do was to place *apple* or *book* following *give*, a sign he readily made when he wanted something. Questions such as *Nim want X?*, *Nim eat X?*, and so on appeared frequently in transcripts of videotapes of Nim's teachers signing with him. What seemed at first to be strong evidence of grammatical knowledge was undermined by this discovery.

I must therefore conclude—though reluctantly—that until it is possible to defeat *all* plausible explanations short of the intellectual capacity to arrange words according to a grammatical rule, it would be premature to conclude that a chimpanzee's combinations show the same structure evident in the sentences of a child. The fact that Nim's utterances were less spontaneous and less original than those of a child and that his utterances did not become longer, both as he learned new signs and as he acquired more experience in using sign language, suggests that much of the structure and meaning of his combinations was determined, or at least suggested, by the utterances of his teachers.

This is not to say that a chimpanzee is simply not capable of creating a sentence. I will elaborate in Chapter 14 my belief that Nim's linguistic achievements may not fully reflect his linguistic potential. It would be foolhardy to ignore what may have been the major obstacle to his further linguistic progress: the emotional turmoil caused by the too frequent departure of teachers to whom he had become attached. But it would be equally foolhardy to overlook the differences between Nim's linguistic development and that of a normal child learning to speak.

14

Beyond Project Nim

Given Nim's accomplishments under circumstances that were far from ideal, it is tempting to think that another project, which could benefit from what was learned from Nim, could produce a chimpanzee with an even greater understanding of sign language. After all, prior to the Gardners' and Premacks' pioneering studies, conventional wisdom was that chimpanzees were not capable of mastering even a simple vocabulary of arbitrary words. If the difficulties handicapping Project Nim could be avoided in a new project, should it not be possible to teach a chimpanzee more about sign language than we were able to teach Nim?

Before I try to answer that question, I will specify two aspects of linguistic development that must occur if a chimpanzee's use of sign language is to surpass Nim's. One has to do with the length of the chimpanzee's utterances, the other with his or her motivation to sign. Despite the steady increase in Nim's vocabulary, the length of his utterances did not increase proportionately. As noted in Chapter 13, during the last twenty months that Nim lived in New York, the average length of his utterances stabilized at around 1.5 signs, in sharp contrast to the way a child's utterances continue to increase in length, variety, and complexity. The question should be rephrased: Is a chimpanzee capable of making utterances whose average length is longer than 1.5 signs?

This question has little to do with the size of the chimpanzee's vocabulary. Of course, a minimal vocabulary is needed before one can obtain a corpus containing enough types of combinations to show whether those combinations obeyed some structural rule. But once Nim acquired a vocabulary of about 50 signs, he generated a large enough corpus of combinations to study. Unfortunately, his utterances had stopped growing in length by the time he had learned about 40 of his 125 signs. Clearly, the sheer size of his vocabulary was not a decisive factor. It would be comforting to think that it was—I feel confident that Nim's vocabulary could have been doubled and perhaps tripled without much difficulty.

I believe that it was Nim's motivation to sign, or his lack of it, that provided the biggest obstacle to the lengthening of his utterances. With few exceptions, it did not appear that Nim had discovered the "power of the word" in the same sense that a child does. The exceptions all concerned one function of language: that of *demanding* something. *Hug, play, dirty, sleep, eat,* and so on were all requests that something happen to Nim. Occasionally, but with fairly low frequency, Nim signed spontaneously to *name* things, for example, *flower, hat, drink* (the last while watching the bus driver pour some coffee), and a host of other signs while thumbing through books or magazines. As far as I could determine, however, Nim would only sign in this manner when he was with people he liked and trusted.

Those teachers constituted too small a fraction of the large group of caretakers who looked after Nim. I believe that it was this unstable aspect of Nim's life that weakened the motivation needed to sign about things other than requests. So the question of whether or not a chimpanzee can learn to use sign language at a level significantly greater than Nim did should really be rephrased as follows: Can one instill a greater motivation to sign than we managed to instill in Nim?

One way to do it, even in a project as limited in resources as Project Nim, would be to start emphasizing the importance of signing at a much earlier age than I did with Nim. The point is not to make a chimpanzee into a signing prodigy but to establish a strong habit of signing before the chimp develops a physical mastery of its world. Instead of allowing an infant chimpanzee to operate on its environment directly by reaching for and grabbing things, I would encourage it to sign about its needs.

The ease with which a child learns language may be less a consequence of superior intellectual machinery than of a child's willingness to inhibit its impulses to grab and to use words instead. In contrast to a child, a chimpanzee seems less disposed to inhibit its impulses, preferring to operate upon the world in a physical, as opposed to a verbal, manner. To get a chimpanzee into the habit of signing, it would help to begin instruction in sign language at an age at which its physical coordination is limited. During the chimpanzee's first year, it is essentially as helpless as a human infant: its locomotion is rather limited, and it is quite uncoordinated in its attempts to grasp things. But during its first year a chimpanzee can sign.

When I watched Nim learn his first sign at the tender age of four months, I was relieved at the ease with which he could be taught to sign. But at that time, I was not aware of the importance of making signing a strong habit that would resist competition from other natural motor habits. While Nim was quite helpless, I should have required him to sign, or at least attempt to sign, for anything he wanted. While such pressure to sign might not have increased the size of his vocabulary, it would probably have strengthened his disposition to sign about things he thought were important. As successful as we were in socializing Nim by getting him to inhibit various actions, I believe that we would have had greater success if Nim had understood at an early age that signing and not grabbing was the way to satisfy his impulses.

Encouraging an infant chimpanzee to sign at the earliest possible age may dispose it to sign rather than to reach for things, but in order to motivate a chimpanzee to sign *about* things, rather than merely demand them, I think it is essential to develop a close and stable relationship between the infant chimpanzee and its caretakers. The difficulties in arranging for such stability remains, in my opinion, the biggest single obstacle to maximizing the linguistic potential of a chimpanzee.

During its first year, a chimpanzee requires relatively little attention. It sleeps a good part of the day, and because of its limited motor ability, it can do relatively little damage. If one wants to teach a chimpanzee to use language, however, it becomes necessary to devote greater attention to everything the chimpanzee does as it gets older. Ideally, that attention should come from the same group of caretakers who have raised the chimpanzee from birth and upon whom the chimpanzee has grown dependent.

Once the chimpanzee is ambulatory, the draining nature of working with it would necessitate a sizable group of people who would take turns keeping it company during all of its waking hours. By the age of two, Nim slept 8 to 10 hours every night, which meant that he had to be looked after approximately 16 hours a day, seven days a week. For this work I would recommend a group of at least eight caretakers. This may seem extravagant for only 112 hours of work a week, but that is actually a gross underestimate of the actual time needed to provide effective teaching. In the first place, that figure makes no allowance for the detailed preparations that a teacher must make before spending time with the chimp. It also makes no allowance for writing the careful and detailed reports without which there would be no documentation of the chimpanzee's use of sign language. These considerations would double the time required of each caretaker for a session and yield an aggregate need for 244 hours a week, which works out to more than six full-time positions.

Allowance also has to be made for the tremendous investment of emotional and physical energy required to engage a chimpanzee and to command its attention so that it can be motivated to learn to sign. Following a good session with Nim, I felt drained for hours. So did the other teachers. During that time, it was not possible to work effectively on other aspects of the project. Finally, since the project cannot stop while a teacher goes on a much needed vacation and untrained substitute caretakers do more harm than good, each vacation means more work for the remaining teachers. The same considerations apply when a teacher needs time off because of illness.

In calling for eight full-time caretakers, I am not taking into account the additional work entailed in data analysis, which requires another small group of workers to tabulate the chimp's utterances, transcribe videotapes, and perform the many analyses necessary to evaluate just what he was learning about sign language.

The costs of the ideal project I have been describing are staggering. A salary of $15,000 a year for each full-time caretaker seems to be the minimum needed to attract a person who is a fluent signer and who has the intelligence, research skill, and personality to be an effective teacher. Adding the going rate of 20 percent for fringe benefits yields a yearly budget of $128,000 just for caretakers. Salaries for data analysts, and a budget for food, clothing, toys, camera and video equipment, film, video-tape, photographic processing, and so on would require at least another $30,000 a year. Here I am assuming that a good portion of the data analysis would be done by graduate students and volunteers. Finally, one has to consider where the chimp and its caretakers would live. I cannot assume that a new project director would be lucky enough to come across a vacant mansion and estate a mere twenty minutes from campus that would be made available for a token rent that barely paid the heating and the utility bills.

Since it is hard to predict what the commercial rent would be for a place that provided the space and privacy we had at Delafield, I will omit housing from the budget I am suggesting for a project that *might* produce a chimpanzee whose level of signing would surpass Nim's. Assuming only modest cost-of-living increases in salary, and not including money for rent, that budget would exceed $1,000,000 in direct costs over a five-year period. Quite a large investment to get one chimpanzee to learn an undefined measure of sign language!

Before we ask whether it would be worth spending that much money on a single chimpanzee, whose longevity cannot be guaranteed for the duration of this hypothetical project, consider the differences between my idealized budget and the one I actually worked with on Project Nim. All told, during the almost four-year span of Project Nim, I received just under $200,000 in grant money. I supplemented this with more than $45,000 of my own money. Would it make sense to quadruple this total for a project that could not be guaranteed to surpass the accomplishments of Project Nim?

The main reason for spending such a large sum would be to insure the presence of a stable group of teachers throughout the project period: preferably eight full-time equivalents of Annie Sullivan, the patient, re-sourceful, and energetic teacher who taught Helen Keller to speak. Without the intensive and intimate bonds that only a stable group of caretakers could create and maintain, I doubt if there would be enough social motivation for a chimpanzee to share its world with its caretakers by signing about it. It may be tempting to try to save the expensive salaries that eight teachers would require by working with caged chimpanzees and using food and drink rewards instead of the love and praise that a teacher can provide. But I do not believe it would produce a level of communication beyond that needed for the chimpanzee to solve the problems it would be required to solve in order to acquire the food and drink it desired. Such communication is considerably more primitive than human language. Indeed, it is hardly necessary to use chimpanzees to show that an animal can learn the "names" of various items of food and drink. Dogs, pigeons, and rats may not be able to learn as many names or learn them as rapidly as a chimpanzee, but there exists ample evidence that they can learn to communicate their desires through arbitrary symbols.

Considering the cost of taking the next step beyond Project Nim, I found myself asking the question put to me by the National Science Foundation when I applied for funds: Would it be worth more than $1,000,000 spent over a five-year period to advance the level of sign language in a chimpanzee beyond that demonstrated in Nim? Although I have no desire to be the recipient of such a grant, I do believe that it would be worthwhile to attempt to teach sign language to not one but a number of chimpanzees under conditions more ideal than those that Nim experienced. Communication with another species at the level of human language would be as exhilarating as receiving a message from outer space.

For a fraction of the cost that has already been invested in trying to communicate with intelligent beings in outer space or to simulate intelligence by programming computers to act in accordance with rules of "artificial intelligence," it may be possible to learn how another natural species views and thinks about our world. Prosaic as a chimpanzee may seem when compared with an imagined creature from another world, we have in fact made considerable progress in showing that it can master some aspects of human language, and we have much to learn about its mind and its world, both radically different from our own. In view of the narrow biological separation between the human and the chimpanzee it seems that a chimpanzee has at least as much potential to reveal primitive patterns of human thought as does a computer.

Even if it cost many millions of dollars to produce a chimpanzee who truly understood the power of signs as a means of communicating about its world, the advancement of knowledge that would result would be well worth it. That accomplishment would place us on the threshold of creating a community of chimpanzees who could sign with one another. It is hard to imagine a more exciting voyage back in time than such a community would provide. The opportunity to observe how the addition of language, as we know it, would influence the culture of a group of chimpanzees might provide a priceless glimpse of what life was like at the dawn of human civilization.

There remain, I am sure, skeptics who doubt that a chimpanzee's knowledge of sign language can advance much beyond that demonstrated in Nim and in other chimpanzees. While I have no way of proving that further progress is possible, the progress made in less than twenty years cannot be ignored. Because I cannot overlook what Nim learned about sign language under conditions that were far from ideal, I feel confident that Nim's impressive achievements will not prove to be the last word.

Epilogue:
One Year Later

During the year that followed his return to Oklahoma, hardly a day went by when I wasn't reminded of Nim. This was not just because of my almost daily efforts to interpret and write about the data and the project in general. I often simply missed Nim and wondered about his adjustment at the Institute for Primate Studies. From regular bulletins about Nim's progress I learned that Nim had recovered from the depression he had experienced following his arrival. Within a month he began to accept his new diet of primate chow and to form close bonds with a number of other chimpanzees, particularly Onan, a full brother, who was a year and a half older than Nim.

I had no reason to doubt these generally encouraging reports, yet I was eager to visit Nim and find out for myself how he was, how he would remember me, and how he would sign. On September 27, 1978, a year and two days after Nim left New York, I flew to Oklahoma to spend a day with him. I had just finished writing the preceding chapters of this book and the first draft of a lengthy scientific article that summarized Nim's knowledge of sign language. After such labors I felt more than ready to see again—and perhaps "converse" with—that wonderful creature who had been a central figure of my existence from the day he was born.

During most of his first year in Oklahoma, Nim lived on a man-made island in the company of about nine other chimps. A reunion on the island would have been possible, but I ruled it out because I wanted the visit to be free from distraction. Besides, I did not want Nim to have an opportunity to see me until he could be right next to me. The arrangement I preferred was for someone to take Nim for a walk during which he would unexpectedly meet me. Dr. Lemmon thought this was a feasible plan and arranged to have Nim taken off the island on the day before my visit. While he was away from the island, Nim would be housed with some of his playmates in a large cage, in the building where he had spent his first night at the institute. Nim was quite familiar with that cage. It had been his home during his first few months at the institute and later whenever it was too cold to be outdoors.

Early on the morning after I arrived in Norman, I discussed my plan for meeting Nim with Alyse Moore, one of the chimpanzee handlers at the institute, and a television crew from a network news show who wanted to document Nim's reaction to me as a general interest story. Alyse would lead Nim to a secluded area across from the chimp island. I would be sitting behind a tree, out of Nim's view. Also hidden behind the tree would be a bag of objects that Nim used to sign about: a picture book, a hat, a ball, sunglasses, and various kinds of food and drink. The television crew would be fully visible but off to one side. Nim would undoubtedly notice them, which made it less likely that he would notice me.

Nim was quite calm as he walked toward the area in which we were to meet. Holding Alyse's hand, he studied the unfamiliar faces of the TV crew. As soon as I turned to greet him and he realized who I was, Nim's mood changed instantly and dramatically. He let out a wonderful shriek, leaped into my arms, signed *hug*, and gave me a tremendous chimp embrace. It was wonderfully familiar. If I had any doubts about his remembering me or whether his reaction would be cool, they were quickly dispelled.

A few hugs later Nim pulled away and began to groom me. When I put him down for a moment he looked a bit puzzled and troubled. After shaking his hands back and forth a few times he again signed *hug* and came back into my arms. Finally he sat down right next to me and gently but firmly ripped open the buttons of my shirt. For a second or two he inspected the wire that led from the miniature microphone I was wearing to the transmitter in my back pocket. Nim then began to groom me systematically as he had done so often in the past. He started by sifting slowly through the hairs on my chest, as if looking for a foreign body he could pull off. He then went through the same routine with my eyebrows and the hair on my head. Finally he pulled up the cuff of my trousers, pulled down my sock, and inspected the lower part of my calf. There he noticed a small scratch. He signed *groom* and carefully rubbed the scratch as if to soothe it. A moment later he continued the familiar routine. He signed *shoe*, then *out shoe* and pointed to my shoe. When I pulled off one of my shoes, Nim picked it up, looked inside it, and promptly put it on his foot. With his foot in my shoe, Nim resumed grooming my foot and tried to pull off my sock. (I had often wondered about Nim's curiosity about shoes and socks. I could understand his interest in wearing my shoes as an expression of his identification with me. But why his interest in bare feet? My hunch is that Nim could never

Reunion with the author

understand why anyone would want to cover such a vital part of his
body. Nim used his feet regularly to grab things and, on many occasions,
to sign. Since chimpanzees often walk on all fours, they probably think of
hands and feet as more interchangeable than do humans. If Nim could
tell me why he routinely tried to remove his teachers' shoes and socks, I
suspect that he would say that he wanted his teachers to be like
chimpanzees.)

Normally I would have allowed Nim to continue grooming me for as long as he wanted, and to make him more relaxed, I would have also groomed him. But I wasn't sure how long Nim would remain attentive to me, and I wanted to find out what he remembered of sign language. Reaching for the bag, the first thing I pulled out was a hat. Nim looked at it somewhat impassively, then signed *give*. I sensed that he was less concerned with the things I brought than he was with me. Nim had done very little signing during the past year. Why, he must have wondered, did I turn to quizzing him on his signs so soon after I saw him? Still holding the hat, I pointed to it and signed *What's that?* Nim slowly made the *hat* sign, and I gave him the hat. I went through a similar routine with a tennis ball. Again Nim did not seem terribly interested in what I offered him. When he did sign *ball* I felt that it was more to please me than to get the ball.

His interest in an old picture book was considerably greater. When I showed him the book he looked delighted and made the quiet "oooh" sound that I had often heard him make when he was relaxed and interested in something. Without hesitation, and without my even asking him to identify the pictures, he signed *toothbrush, hat,* and *dog* to the appropriate pictures. When I put the book away, Nim did not seem terribly

Nim has not forgotten how to sign *hat.*

disappointed. With a beseeching look he gazed at me steadily with his lower lip pulled down from his upper lip. He seemed to be telling me that he was glad to see me. Out of habit, I had secured the end of his lead under my foot. But it was plain that this was one time he would not try to run away.

Were it not for my curiosity about his signing and the pressure from the camera crew, who wanted to film as many of Nim's signs as possible, I would have been happy to simply sit with him and allow him whatever time he needed to get used to me again. I decided to have breakfast with him. That would certainly produce a lot of signing and make Nim feel as secure as possible about my presence. No sooner had I pulled out a bottle of orange juice than Nim signed *me Nim drink* and *drink me*. After he drank his first cup he signed *more drink*. He signed *more* while holding the cup in his mouth and *drink* while holding the cup in his hand. During breakfast Nim made many signs, each appropriate to the food or drink he wanted: *apple, orange, grape, banana, yogurt,* and *cookie*. These signs were often combined with *eat, drink, more, me,* and *Nim*. Nim signed *apple, orange, grape,* and *banana* spontaneously, as soon as I showed him these objects.

During the half hour I sat with Nim in front of the TV camera, he made nineteen different signs. Undoubtedly, he would have made many more if I had signed with him in a play situation or if I had brought other

He still remembers the sign for *orange*.

objects (for example, crayons of different colors, matches, or pictures of his teachers or of other foods). I felt that Nim's signing was a bit rusty but that with a little practice he could become as proficient as he had been when he left New York. I was later told that only a few isolated attempts had been made to sign with him and that no one had seen him sign with such frequency and precision as he had on that morning.

When the TV crew was finished, I relaxed and stopped trying to get Nim to sign. Nim responded by coming closer to me and gazing intently at my face. A few times he tried to groom my eyebrows. On the whole he seemed content to sit next to me and renew our long acquaintance.

I had hoped to have an opportunity to visit with Nim for a few days, in order to get a better picture of his growth and development since I had last seen him, but my teaching schedule barely allowed me the time to stay for a single morning. Yet it was long enough to prove that the year he had spent in the almost exclusive company of other chimpanzees had not seriously weakened his attachment to me. He probably would have reacted in the same way to Stephanie, Laura, Bill, Joyce, Dick, and other teachers with whom he was close. No doubt they would have been as thrilled as I was to see first-hand that, even though Nim was a bit bigger, he was just as sweet and delightful to be with as before.

Nim was not scheduled to return to the island until late that afternoon. In the meantime he was to stay in the building in which he had slept the previous night. Instead of handing Nim over to Alyse, I decided to walk with him to his cage and observe how he got along with its other occupants: Onan, his six-and-a-half-year-old brother, and Lilly, who was also six. Nim did not put up much resistance to entering the cage. During the few moments I fidgeted with his collar in order to disconnect his lead, he looked at me somewhat plaintively, perhaps dimly aware that I was soon going to leave. Nim lingered at the door of the cage while I stood next to him. When I pointed to the inside of the cage he slowly turned away from me and offered no resistance to my closing the door.

Once inside the cage Nim stared at me intently, oblivious to the presence of Onan and Lilly. It was clear that Nim was not very happy about being separated from me once again. But it was also clear that being placed in a cage was not nearly as upsetting to him as it had been a year earlier. This time there were no screams, tantrums, or violent thrashing around. Instead, Nim watched me quietly, as if he had no hope that he could return to the life I had provided for him during his first four years.

After I took a few steps back from the cage Nim turned to Onan, who responded by engaging him in a brief bout of play-fighting. The big grins on their faces made it obvious that this was a pleasant tussle and that there was no cause for alarm. As quickly as Onan and Nim started wrestling with each other they stopped. Nim then went to another corner of the cage, where Lilly was caring for an infant chimpanzee whose mother had recently died. Lilly paced around her section of the cage clutching the adopted infant to her chest. Whenever Nim approached she shooed him away by flicking her hand in his face. Each time this happened Nim backed off. Previously I had seen Nim act the same way toward small children whose parents didn't want them to play with Nim. In those situations Nim was interested both in playing with the child and in asserting his dominance. He was also wary of being punished by an adult if he got too rough. So it was with Lilly's adopted baby. When Lilly put her baby down, Nim cautiously approached with his eyes directed, not at the baby, but at Lilly. Lilly sat nearby, ready to intercede if necessary. When Nim got close enough he jabbed his finger toward the infant and just managed to touch its shoulder. Almost before he finished this parry he jumped back in the opposite direction as if he anticipated that Lilly would let him know what she thought of his games.

Various staff members at the institute had told me that Nim was even more daring and aggressive on the island, and the consensus was that within a year or two he would emerge as one of the dominant chimps of the colony. What he lacked in size he more than compensated for in cunning. After a mere five months on the island, only three of its nine inhabitants ranked above Nim in the dominance hierarchy. Two of the chimpanzees whom Nim dominated were at least a year older than he was. After watching Nim interact with the other chimpanzees, I was confident that he had made a good adjustment to his new companions and that his progress in dominating other chimpanzees at the institute would continue unabated. That thought, however, did not make it any easier for me to leave. I was sure that he would continue to do well without me, but I, on the other hand, would always miss Nim.

I waved goodby to him. He looked up at me, raised his hand as if to make a weakly executed *bye* sign, and then resumed his antics with Lilly and her baby.

Appendix A:
Sign Language

Talking with one's hands is often regarded as undignified. Yet people would find it hard to converse without using some hand movements and facial expressions to emphasize what they have to say. In many situations, we talk with our hands alone. Consider the following list of obvious examples:

quiet (finger to lips)   signals in sports contests (e.g.,
over there (pointing)   time out, penalties, foul,
you   touchdown, ball, strike)
stop   bidding at auctions
clenched fist   V (for victory)
throwing a kiss

People who find themselves in a foreign country are often able to make themselves understood through a frenzied combination of gestures. Communication is possible even without a word of spoken language. Facial expressions and body language assume a more stylized form in the arts of pantomime and charades. People derive immense pleasure from watching mimes such as Marcel Marceau, and have no trouble understanding their "stories."

Unlike pantomime, charades, and everyday gestures, sign language is a natural language. It is as valid a medium of communication as any spoken language, even though the medium of communication is *visual* rather than auditory. Just as the practitioner of a spoken language must learn to discriminate among the different combinations of sounds and intonations that define different words, the user of a sign language must learn to distinguish between the different combinations of specific movements, hand shapes, facial expressions, and body postures that define different signs. Through the subtle and complex use of the space in front of the body, movements of a sign, and repetition, a skilled signer can encode any of the meanings and relationships that speakers of spoken language can encode through inflections, suffixes, participles, gerunds, and other grammatical devices.

No one is really sure just when deaf people began to communicate with one another through signs. References to sign language can be found in the Bible. It is known that the use of sign language was widespread in America (as well as in other countries) during the eighteenth century. It is not clear, however, what was the specific nature of these sign languages. Traditionally, deaf people were discouraged from using signs. Those who did were regarded as insane and often placed in mental institutions because of their inability to communicate through spoken language. The community of speaking people tried, and *still* tries, to get deaf people to use spoken language. Deaf people have been urged to read lips and, through great effort, to manufacture the sounds of spoken language—sounds they cannot hear.

It was not until the late eighteenth century that a Frenchman, the Abbé de l'Epée, began a movement that ran contrary to a long-standing prejudice against sign language. He recognized the spontaneous efforts of deaf people to sign as a natural language. Instead of encouraging them to communicate as hearing people might, he sought to standardize the signs he observed various deaf people use. He invented new signs for certain inflections and for concepts of spoken French that seemed to be absent from the vocabularies of the deaf people he studied. De l'Epée's achievements spread to America through the work of Thomas Hopkins Gallaudet, an American minister who went to Paris in 1817 in search of a method for educating deaf people.

The gestures of a traffic cop, a sports official, a minister, and a mime are related only superficially to the formal signs of sign language. It is true that gestures are used in all of these examples of nonverbal communication. It is also true that some of the gestures of sign language still reveal their origins in the transparent language of hand movements and mime. More often, however, sign language assigns meanings to gestures in an arbitrary manner. It is rarely possible to infer the meaning of a sign from its form. In this respect, sign language is more similar to spoken language than it is to mime and gestural forms of expression.

An uninformed observer would have difficulty understanding most of the signs of sign language. In many cases, universal gestures (such as *stop*) are expressed in sign language in a form that most people would not recognize. The arbitrary nature of sign language is revealed further when one considers how many forms it takes. There is no universal sign language. Fluent signers of Chinese, Israeli, British, French, Swedish, American, and Japanese sign language (to name only the best-documented sign languages) would not be able to understand each other without the aid of sign interpreters. And countries that speak the same language, such as the United States and Great Britain, do not necessarily use the same sign language. American Sign Language (ASL) has much more in common with French sign language, to which it is related historically, than it does with British sign language.

stop

It has been estimated that ASL is used currently by more than half a million deaf people in North America (as well as by thousands of hearing people who interact with the deaf). This makes ASL the fourth most widely used language in America. (English is first; Spanish, second; Italian, third.)

A relatively small percentage of signs of American Sign Language (ASL) contain elements of their referent. For example, some sources attribute the original forms of the signs *man* and *woman* to the different types of hats worn by members of each sex. Men wore top hats and women bonnets. At the turn of the century, the sign for man was made by moving the hand away from the forehead, the thumb touching the index and middle fingers, as if to touch the brim of an imaginary top hat. The sign for woman was made by moving the fist, with the thumb slightly extended, away from the cheek, as if to pull on an imaginary bonnet string. With the passage of time, the original movements were dropped and these signs were made simply by touching the forehead in the case of *man*, and the cheek in the case of *woman*. Many other signs appear to have evolved from earlier "iconic" representations of their referents into more abstract forms. An iconic sign is a sign whose meaning can be inferred from its form.

American Sign Language has a vocabulary of approximately five thousand different signs. This figure, however, is deceptively small because it does not take into account how the meaning of a single sign can be changed systematically through the use of movement, repetitions, and physical space. The structured use of facial expressions, body shifts, and eye gaze (integral parts of the grammar of ASL) allows the signer to add additional nuances of meaning. Through these devices, it is possible to convey a much greater range of meaning than would seem possible with a vocabulary of approximately five thousand basic signs. *Forest*, for example, is signed by repeating the sign *tree* in different spatial positions. Likewise, *city* is signed by repeating the sign *house*. *Village*

tree |———————————— *forest* ————————————|

|——————— *house* ———————|

|——————— *city* ———————|

is signed the same way but with fewer repetitions.* These examples should
make clear that many signs represent concepts rather than specific words.
They also illustrate how the space in front of the signer's body is used to
encode grammatical meanings.

The easiest signs for a nonsigner to understand are iconic signs. Shown
below are photographs of mime artist Jim Moore communicating the meanings
of some iconic and noniconic signs. The meanings of his gestures are quite
transparent.

"boy"   "girl"

"airplane"   "stop"   "think"

* See the footnote on page 177 for a description of the role of repetition in
grammatical contrasts, for example, in differentiating verbs and nouns.

SOME EXAMPLES OF "ICONIC" SIGNS

| *hug* | *pray* | *think* |

SOME EXAMPLES OF "NONICONIC" SIGNS

| *girl* | *boy* | *airplane* |

All signs, iconic or otherwise, are made by holding the hand(s) in a particular configuration, and by moving the hand(s) so they touch a particular location on the body with a particular orientation. A major step toward providing an objective and systematic description of signs was the development of a system that identified the basic elements of sign language. William Stokoe, a linguist at Gallaudet College, in Washington, D.C. (the only liberal arts college for deaf people in the world), has shown that every sign can be described as a unique combination of nineteen hand configurations, twenty-four types of movement, and twelve body locations. This system is analogous to the phonemic system for specifying the words of spoken language. A simple illustration of Stokoe's system of cheremes (the manual equivalent of spoken phonemes) can be seen below and on the next page.

| ⊢——————— *flower* ———————⊣ | ⊢——————— *kiss* ———————⊣ |

|——————— home ———————| |——————— yesterday ———————|

Flower, kiss, and *home* are made with the same hand configuration, a "tapered *o*-hand" (see pages 241–42 for illustrations of how different letters are signed), and with the same side-to-side movement across the face. The only thing that distinguishes these signs is the location at which the hand touches the face. The cheremic distinction between *flower, kiss,* and *home* in sign language is similar to the phonemic distinction between the spoken words *mill, kill,* and *will.* The change of one element of the series of sounds that defines these words makes for an important change of meaning. (*Yesterday* is made with a different hand configuration from that used to sign *flower, kiss,* and *home*—the "*y*-hand"—but the hand touches the cheek in the same location as for *home.* Thus, *home* and *yesterday* differ only with respect to hand configuration.

The similarity between Stokoe's cheremes and the phonemes of spoken language was revealed in an unexpected way in studies of memory. When human subjects learn a list of words, their errors are usually based on the sound of the words and not their meanings. Thus, when asked to remember words such as *mill, bat,* and *tan,* errors are more likely to include words that have similar sounds (such as *till, pat,* and *pan*) than words that have similar meanings (such as *factory, animal, bird,* or *brown*). The phenomenon of substituting words that have a similar sound when trying to remember a particular word has an analog in sign language. When trying to remember a list of signs such as *flower, yesterday,* and *man,* subjects are more likely to substitute *kiss, home,* and *think,* signs that are topographically related to the signs the subject is trying to recall, than semantic equivalents such as *leaf, previously,* or *boy.*

Until recently, sign languages have had no written form of their own. Instead they have had to rely on the written versions of the languages of their culture. The absence of a written mode of expression is more an inconvenience than a fundamental weakness of sign language. Spoken language existed for many thousands of years before written forms were invented. Indeed there are still many spoken languages for which written forms do not exist. Even though word of mouth may not always prove as efficient as writing for communicating certain kinds of information, we should not lose sight of the fact that spoken language remains man's basic form of communication. Sign languages have had no need to invent written forms of their own because they have existed in cultures that have written forms of the spoken languages. If those written forms were not available, it would be quite easy to invent a system of symbols that could convey all of the information conveyed by the expressions of sign language.

The relatively small vocabulary of sign language could prove to be a more serious limitation. Even though the range of meanings that can be expressed in sign language can be expanded by the spatial devices I described earlier, there are no signs for the names of people, certain technical terms, and certain places. Sign language solves this problem in a number of ways. The word is finger-spelled, a new sign is invented, or the intended meaning is conveyed by some variation of a root sign. As described below, these variations are the essence of sign language.

A finger-spelled word is a *coded* representation of that word. Just as Morse code represents each letter of the English language by a unique combination of dots and dashes, finger-spelling provides a unique hand configuration for each letter. As such, finger-spelling is not a language. It provides nothing more than a way of representing the words of a language. In some cases, however, ASL adopts the configurations used in finger-spelling in much the same manner that English adopts Greek and Latin roots to form new words.

The manual alphabet of finger-spelling (as used by a right-handed signer) is shown below and on the next page. Most letters are signed with the hand in a stationary position. The exceptions are *j* and *z*, both of which call for "drawing" the shape of the letter in the air, with the hand in the correct configuration. In finger-spelling a word in sign language, the signer uses the English spelling of the word. Fluent signers do not enunciate every letter. The movements required to shift from one letter to another convey enough information for recognition.

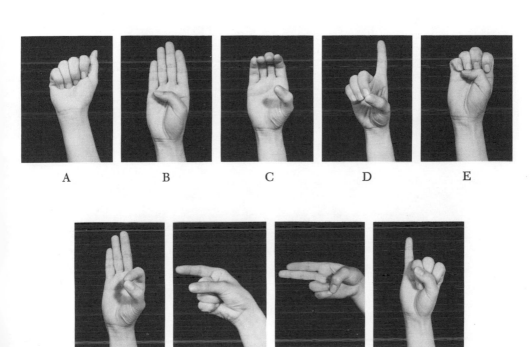

A   B   C   D   E

F   G   H   I

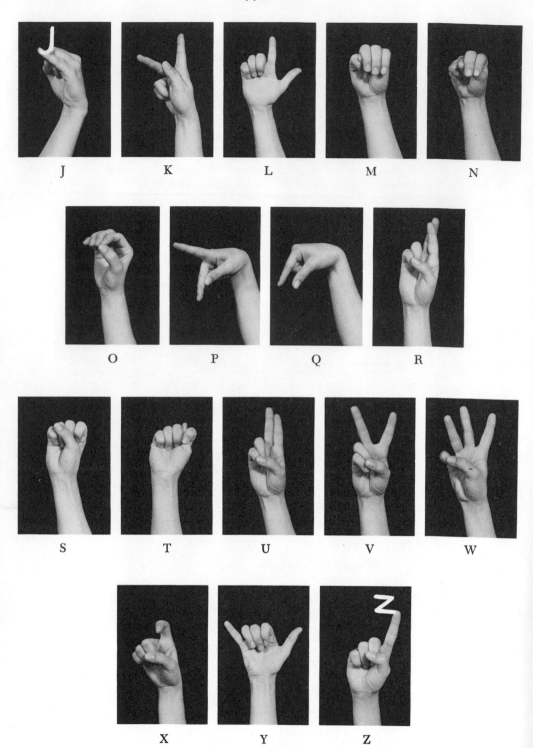

J K L M N

O P Q R

S T U V W

X Y Z

Finger-spelling *protein*, a word that is not now represented in the vocabulary of ASL. A fluent signer, of course, would not need to "freeze" in each position as Laura is doing here.

P   R   O

T   E   I   N

Signs that are invented are often part of a dialect used by a small group of signers. Perhaps the most common stimulus for inventing a sign is the need to identify individuals. Name signs often capture some feature or characteristic of the person named. The configuration of the hand used in signing someone's name sign is often the manual alphabet form of the first letter of that person's name. Consider some of the name signs of Nim's teachers. Laura, Susan, Dick, and I each used the first letters of our names. Laura's name sign was made by holding the *l*-hand with the thumb touching the eyebrow. As is the case with many name signs, a number of factors contributed to the invention of her sign.

Some of Nim's teachers sign their own names.

Laura   *Susan*   *Dick*   *Herb*

At the project meeting in which Laura's name sign was invented she had just finished describing a recent trip to Europe and her love for traveling. This suggested to me that her name sign should be a variation of the sign for airplane (see page 243). Another project member pointed out that Laura had a habit of brushing her hair away from the lower part of her forehead. Since the sign for the first letter of Laura's name is contained in the sign for flying, I suggested that her name sign be the letter *l* touched to one eyebrow and then brushed back over her hair. Neither Susan nor Dick was able to incorporate any special personal characteristics in devising their name signs. Susan signed her name by crossing her right hand over her chest and touching the base of her left shoulder. The right hand was held in the *s* configuration. In Dick's case, we felt that a *d*-hand (touching the thumb to the third finger with the index finger extended) would be too hard for Nim to make, so we simplified it as shown. My name sign was made by touching the *h*-hand to the base of my mustache.

Signed English, a form of sign language that models itself specifically on spoken English, regularly incorporates letter signs to distinguish words. For example, in ASL there is only one sign for *group, class, organization, family,* and *society*; they are distinguished by the context in which they are signed as well as by facial and body expressions and other syntactic devices. In signed English, each of these words is signed with the hand in the configuration of the initial letter (*g, c, o, f,* or *s*).

Signed English uses letter configurations to distinguish between words that are not differentiated in ASL

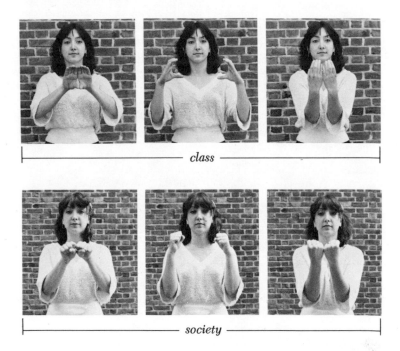

class

society

family

group

organization

Signs whose hand configurations are based upon the first letter of the written form of that sign can be found even among standard signs of ASL. The days of the week are signed by holding the right hand, facing the listener, in the configuration of each day and rotating the hand clockwise. Thursday is distinguished from Tuesday by switching from the *t*-hand to the *h*-hand during rotation. Sunday is an exception to this rule. It is signed with both hands and the configuration of each hand is the same: an open hand, palm out. According to some sources, this configuration, which is not represented in the manual alphabet, is modeled after a minister performing a service.

The names of certain colors (green, yellow, blue, violet) are also made by jiggling the right hand clockwise, in the configuration of the first letter of the color. Colors such as red, black, and orange are signed differently. *Red* (compare below) is signed by touching the index finger to the lower lip. The sign for oranges that one eats is the same as the sign for the color orange: the closed fist is touched to the mouth. *Black* is signed by rubbing the index finger above one's (presumably black) eyebrows.

SIGNING THE DAYS OF THE WEEK

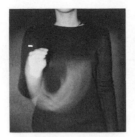 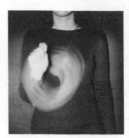 

Monday   *Tuesday*   *Wednesday*

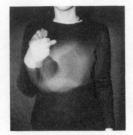 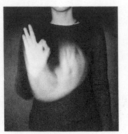 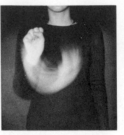 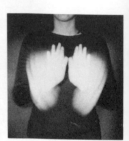

Thursday   *Friday*   *Saturday*   *Sunday*

SIGNING THE NAMES OF COLORS

red   *orange* |———————— *black* ————————|

    ASL is always signed within a spatial framework. This framework and the movement made by the signer to and from various sectors of his signing space provide an efficient means of conveying various grammatical relationships. *Past, future,* and their synonyms are made with respect to a time-line that runs through the body, from one side to the other. A movement toward the back, such as the right hand moving backward over the right shoulder, signifies *past*. Similar movements are involved in the sign *yesterday* (see page 240). A flat vertical right hand moving forward from the right cheek is the general sign for *future*. This sign represents the concept of future as expressed in the English words *will* and *shall*. The sign *tomorrow* is made by moving the right hand forward in the form of a closed fist with the thumb extended from the right side of the face. Notice that the sign *tomorrow* is the opposite of the sign *yesterday*. *Every day* is made by repeating the sign *tomorrow* twice.

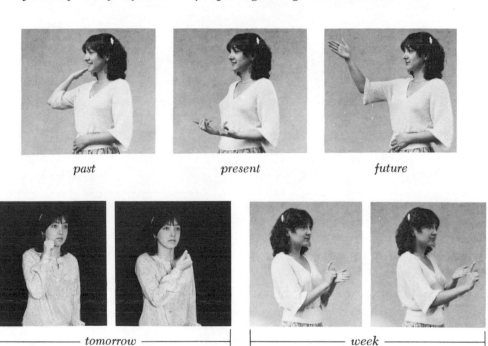

| *past* | *present* | *future* |

| *tomorrow* | *week* |

The sign for *week* is made by brushing the index finger of the right hand outward along the palm of the left hand. For many signers this movement symbolizes moving the hand across one row (week) of a calendar. Below and on the next page we see how *last week*, *next week*, and *two weeks ago* would be expressed in sign language. Each meaning is expressed in one fluid motion that combines elements of the relevant individual signs.

last week (*week* and *past* combined)

next week (*week* and *future* combined)

two weeks ago

Direction of movement within the signer's spatial framework also conveys important information about the meaning of signs. A change in direction often reverses meaning. Antonyms such as *with* and *without*, *join* and *disconnect*, are made with the same hand configurations but with opposite movements. Negation is also often indicated by a simple reversal of movement. The sign for

join

disconnect

without

want can be modified to *don't want* by flipping over the hands as they are drawn toward the body. A similar relation exists between *know* and *don't know*. In addition to the reversal of movement, the signer must also shake his or her head while signing a negative.

want

don't want

know

don't know

Movement can also be used to add subject and object pronouns to various verbs. Sentences such as *I teach you, you teach me, you teach him, he teaches you,* and so on differ only with respect to the direction of movement. These movements are made along a line called the sight line. The sight line refers to an imaginary line that can be drawn from one signer's face to the other's. When the sign *teach* is made along the sight line toward the signer, the signer's eye gaze and the direction of the signer's movement indicate that it is the addressee who is teaching the signer. If the sign originates from the signer and moves along the sight line toward the addressee, that indicates that it is the signer who is teaching the addressee. If the sign originates to the side of

I teach you   *You teach me*

teach

— *He teaches you* —

the sight line between the signer and the addressee, that indicates that a third person is doing the teaching. Who is teaching and who is being taught is communicated by the direction in which the sign *teach* is moved. The same process occurs with other verbs such as *inform, look at,* and *give.*

The passages below will illustrate how certain aspects of meaning are conveyed in sign language.* The reader should regard sign-language "translations" as *minimal* descriptions of what a signer would actually sign and not as a "pidgin English" version of English prose. It is not possible for the written gloss of the original passages to capture all of the nuances of meaning of those passages. Most important aspects of sign language cannot be specified without extensive descriptions of the spatial positions of different signs and the facial expressions of the signer. A *signed* version of the original English passages would convey all of the information and meanings contained in those passages.

English Original

Rain

A little while ago I thought the rain might stop, but now it is raining harder than ever. The wind has changed. Now it is coming from the north, and the temperature is dropping. If it gets much colder, it will turn to snow. I want the rain to go away. It has been raining every day lately, and it is really monotonous. I have been bored stiff. What I want is some blue skies and some warm sunshine. I know we need rain for the sake of the farmers' crops. If there were no rain, everything would dry up and the crops would wither. I have an idea! Why doesn't it rain during the night and stop each morning? Or why doesn't it rain on Fridays and clear up on Saturdays? If I were in charge of the weather, I would see to it that Saturday and Sunday would always be pleasant. On Mondays, who cares? Let it rain.

Sign-Language Gloss

Rain

A-little-while-ago think me rain maybe stop. Now worse. Wind change (pause) now blow from north, and temperature drop. Happen temperature plunge (pause) will begin snow. Me want rain disappear. Every-day rain, rain, rain (pause) truly same, same, same. Me bored. Wish sky blue sunshine warm. True, know rain need for farmer their grow area. Rain disappear, all dry (pause) all wither. Idea! All-night rain, morning stop. Why not? Fridays rain, Saturdays nice. Why not? Me rule, make fine. Me decide Saturday Sunday always pleasant. Mondays (pause) Phooey! Let rain.

* From H. W. Hoemann, *The American Sign Language* (Silver Spring, Md.: National Association of the Deaf, 1976).

English Original

Summer Camp

This year my sixteen-year-old daughter spent her first summer away from home. She was in a work-study camp in North Carolina. I have never been there, but I know about where it is. It is near the South Carolina border, not far from the South Carolina School for the Deaf. My daughter flew down from Ohio at the beginning of summer. When the season was over, she rode with some friends to Washington, D.C., where I had been attending the Seventh World Congress. When she arrived in Washington she was all excited about her experiences at camp. She said her teacher was really strict and forced her to study hard, but I find that hard to believe. I could see that she enjoyed the whole experience. Already she has said she wants to go back next year. We spent two days in Washington, I wanted to see some people at Gallaudet College, and she wanted to do some sightseeing. We were satisfied to limit our stay in Washington to two days. After that we had enough. We were ready to go home.

Sign-Language Gloss

Summer C-A-M-P*

Now year my daughter old sixteen first time summer live away home. Go to C-A-M-P, work study C-A-M-P, in N.C. Me there never, but area where? Me know. Near line S.C. near S.C. institution. C-A-M-P, institution, not far, near. Ago summer begin, daughter fly there from O-H-I-O. C-A-M-P finish, group, her friend, drive Washington, person ride-along. Me there same time. Why? Seventh World Congress. Daughter arrive Washington, truly excited from experiences ago C-A-M-P. She explain, teacher strict, force study, study, hard. Me doubt. Can see, she enjoy all summer. Finish say next year want again go there. We-two stay in Washington two day. Me want see there there there people Gallaudet College, daughter want see there there there around Washington. We satisfied limit two day stay Washington. She finish, me finish, enough. We-two ready go home.

In the first narrative, the hyphens in *A-little-while-ago* suggest that the whole phrase is conveyed with one signing movement. But nothing in the first sentence suggests that *think* is signed with a thoughtful expression or that *maybe* is signed with a tentative expression. Nor is there any indication of how a long English sentence is broken up into short phrases, separated by discrete pauses. Many unessential words are not signed even though signs exist

* Capitalized letters separated by hyphens indicate that these letters are to be finger-spelled.

for them in ASL. For example, "but" in the first sentence of the English original is omitted. Its meaning is conveyed by the short pause between *maybe stop* and *Now worse*. And when the signer signs *Now worse*, he would do so with an air of resignation—a loss of the hope conveyed in the first sentence. The sentence *The wind has changed* is reduced to *Wind change*. That suffices to convey the same meaning as *The wind has changed* for two reasons: the article *the* is redundant; and the context has already placed the action in the past. The sentence beginning *Happen temperature* is an idiomatic construction for conveying a conditional meaning. *Happen* in sign language is the equivalent of the English phrase "if it should happen." The outcome of the drop in temperature (*will begin snow*) is related conditionally to *Happen temperature*, the prior phrase, in two ways: there is a long pause between phrases; and each phrase of this conditional sentence is signed in a different area in front of the signer's body. Thus, the second phrase is separated from the first phrase both in time and space. The meaning of the phrase "it is really monotonous" is conveyed by the signer repeating the sign *same* with a bored look on his face.

Other condensations of the English original include *me bored* for "I have been bored stiff," *Me rule, make fine* for "If I were in charge of the weather," and *all dry* for "everything would dry up." *All dry* is signed over a wide area that helps to convey the idea of many crops. Here again the judicious use of the signer's space helps to convey a nuance that is not apparent from the printed form of the gloss.

The second passage provides additional examples of how English prose is condensed when its meaning is signed in ASL. It also illustrates a number of words that need to be finger-spelled—*camp, Ohio, North Carolina* (N.C.), and *South Carolina* (S.C.). One important aspect of the spatial nature of sign language that is not apparent from the printed gloss is how different places are represented. The best example occurs in describing the signer's daughter's trip from Ohio to Washington. Ohio is signed off to one side of the signer's body; Washington to the other. When signing *drive Washington* and *person ride along*, the signs *drive* and *ride* are made in the direction of where Washington is represented. Later, when the signer and his daughter were ready to leave Washington, the relevant signs were made away from where *Washington* was located in the signer's space. Similar devices are used in signing about relationships between people. The relevant parties are first identified by their sign names and then positioned in the signer's space. Later the signer makes signs, moving from one position (person) to the other, to indicate what is transpiring between the people he is signing about.

The two examples of sign language narrative hardly do justice to the richness of expression that is inherent in sign language. There is really no substitute for seeing a signer convey all sorts of information simply by staking out various ideas in a signing space, by modulations of movement, eye gaze, facial expression, pauses, and other devices that cannot be captured in a printed gloss. Sign language has its own poetry, slips of tongue (hands?), and puns. For the purposes of everyday communication, I can think of no story or idea that cannot be effectively communicated in sign language. It is a rich natural language whose nature is being studied for the first time by psychologists and linguists.

Only recently has sign language been recognized as a valid form of expression that is as effective a medium of communication as any spoken language. One would hope that we will soon see a rapid transition from the stigma that has been attached to deaf people's signing to a full understanding of the richness of an intriguing natural language.

Appendix B:
Recruiting Nim's Teachers

Most of Nim's teachers were recruited from classes at Columbia and other colleges in New York City. An unexpected source was an article in *New York* magazine in February 1975, which brought forth a deluge of more than fifty applicants. As one might expect, the backgrounds of these people were considerably more diverse than the student recruits. However, the eight people I selected from this group had more than ample motivation to learn how to become effective teachers.

My initial encounter with a prospective teacher usually took place in a group interview. Originally, I followed this procedure out of sheer necessity. There was simply not enough time to talk to each applicant on a one-to-one basis. Even when Nim had just turned two, fewer than one applicant in twenty was selected to try out as one of his teachers. That ratio fell to less than one in forty as Nim became older, stronger, and less dependent on his teachers, and as I developed a clearer picture of the type of person who would make a good teacher.

After a few group meetings with prospective teachers, I concluded that the groups were more informative than individual interviews. I was better able to provide a clear and extensive description of the joys and problems of the project when I could reach many people simultaneously. It was also easier to compare how people responded to my questions when I met with them as a group than when I asked the same questions in a series of separate interviews. Another benefit was more subtle. A successful teacher not only had to be effective with Nim but also had to be able to coordinate his or her activities and observations with those of other teachers. In order to do this it was necessary to maintain good relationships with other teachers. Group interviews gave me a sense of how different people interacted with one another and whether they were able to recognize the necessity of a team approach to teaching Nim.

At each group interview I summarized the goals of the project and how much progress had been made toward realizing them. I also informed each group about our financial situation and why everyone would have to start out as a volunteer with only the slim hope that ultimately some of them might land a paid position. Much of what I said was aimed at discouraging all but the most able and motivated applicants. As vividly as I knew how, I described the frustrations of not being able to control or to teach Nim. Apparently, my descriptions of the problems a teacher would encounter were not vivid enough. Virtually every teacher I selected had to discover those difficulties first-hand. Following that discovery, I was often admonished for not emphasizing the difficult nature of the work I had asked for.

I can recall conducting at least eight group interviews to seek new teachers for the project. I also recall my lessening enthusiasm for such meetings as the project evolved. I regarded each successive effort to find new volunteers as a double step backward. Looking for new volunteers was a painful reminder that the project had not reached the point where it could function without volunteers. A more immediate problem was the interruption in the continuity of Nim's upbringing caused by replacing a volunteer, which usually occurred because the volunteer had other obligations. In some instances, however, volunteers had to leave simply because they were not effective with Nim.

Given the goals of the project, the attributes of an ideal volunteer were fairly obvious: a person who knew sign language; who was able physically and psychologically to confront Nim when necessary; who was knowledgeable about linguistics, cognition, and behaviorism; who had taught young children; who was a mature and energetic worker; who had a strong intrinsic interest in the goals of the project; and who would get along well with other volunteers. Of all the applicants to the project, I can recall only one (Ronnie Miller) who I felt came close to approximating my ideal during our initial interview. Clearly, certain compromises were necessary in selecting Nim's teachers.

New volunteers rarely came equipped with a knowledge of sign language and a solid understanding of the intellectual background of the project. But I felt that such knowledge could be taught and that it was more important to select for less tangible and teachable qualities such as motivation, character, and general intellectual ability. Better to select teachers with whom Nim would engage than teachers who were well-versed in psychology and sign language but who could not communicate that knowledge to Nim.

I selected new volunteers by first rejecting certain types of people who,

Nim's Teachers

Teacher's Name	Age When Teacher First Worked with Nim	Student At
Jean Baruch	45	
Michele Baruch	16	Chapin School
Walter Benesch	24	
Fred Bever	12	Cathedral School
Andrea Brin	28	
Joyce Butler	19	Purchase College
Steve Carey	24	Columbia University School of General Studies
Shirley Carse	27	Columbia University Teacher's College
Rina Cascone	29	
Diane DeSotto	19	Ramapo State College of N.J.
Pat Dobro	20	Queens College of New York City

from my point of view, were obviously unsuited for the project. These included people who didn't seem able to see beyond Nim as a cute young infant; people who I felt were attracted to the project as a social center or as a place where they could vent their opinions about language, chimpanzees, or, worse still, unrelated matters; and people who could not make a convincing statement as to why they wanted to work on the project and what they hoped to accomplish. On the positive side I sought an assertive disposition that made me confident that the prospective volunteer could confront Nim and take the initiative in focusing his attention.

A major exception to the rule that new volunteers lacked adequate knowledge of sign language and psychology was Ronnie Miller, a young woman who became interested in the project after reading about it in *New York* magazine. Since both of Ronnie's parents were deaf, ASL was her first language. After completing a major in psychology at NYU, Ronnie had started a successful career in personnel management. Six months after she began to work on the project, Ronnie offered to come to group interviews and evaluate applicants for positions on the project. I welcomed her offer not only because I respected her judgment of people but also because she showed an excellent understanding of the goals of the project. Because of her exceptional work as a teacher and trainer of teachers, I also invited Laura Petitto to group interviews both to assess new volunteers and to talk about her experience as a teacher. At each group interview, Ronnie, Laura, and I rated each applicant. The correlations among our ratings were gratifyingly high, particularly in the "outstanding" and "clearly reject" categories. Given the varied nature of the applicants, it was encouraging to see so strong a consensus.

A summary of the ages and backgrounds of all of Nim's teachers follows.

Occupation	Worked On Project	Taught In Classroom	At Home
	4/75– 3/76	√	√
	6/75– 3/76		
social worker	4/75– 9/76	√	√
	11/74– 4/76	√	
teacher of children with learning disabilities	9/75– 5/76	√	
	6/76– 9/77	√	√
	4/77– 9/77	√	√
	4/75– 2/76	√	
musician	4/74– 6/75		√
	4/74– 9/74		√
	12/74– 3/75	√	

Teacher's Name	Age When Teacher First Worked with Nim	Student At
Renee Falitz	20	
Betty Feinberg	20	Barnard College
Penny Franklin	27	
Connie Garlock	31	Columbia University School of General Studies
Fred Gross	21	C. W. Post College
Tony Harrison	16	Walden High School
Steve Hasday	26	
Joe Hosie	23	Brooklyn College
Maggie Jakobson	14	Calhoun High School
Bob Johnson	25	Columbia University School of General Studies
Alex Kandabarrow	28	Columbia University School of General Studies
Amber Kassman	19	Barnard College
Jaynie Kozai	19	Columbia University
James Kozai	19	Barnard College
Myra Ladenheim	20	Barnard College
Stephanie LaFarge	38	Columbia University Teachers College
Jennie Lee	14	Calhoun High School
Joshua Lee	11	Little Red School House
Andrea Liebert	21	State University of N.Y. at Buffalo
Sue Liebow	19	Barnard College
Tom Martin	23	Empire State University
Sheila McGee	18	Barnard College
Anna Michel	30	
Ronnie Miller	27	
Carolyn Moore	23	Columbia University School of General Studies
Marika Moosbrugger	25	
Dorothy Moscow	53	
Richard Muller	26	
Donna Orloff	30	
Lisa Padden	18	Hood College Maryland
Laura Petitto	19	Ramapo State College of N.J.
Mary Phillips	28	New York University
Susan Quinby	27	

Occupation	Worked On Project	Taught	
		In Classroom	At Home
sign language interpreter	9/76– 8/77	✓	✓
	9/76–11/76	✓	
editor	1/74– 6/76	✓	✓
	12/73– 6/75	✓	✓
	8/77– 9/77		✓
	4/75– 6/76	✓	
clothing promotion	4/75– 6/76		✓
	8/77– 9/77		✓
	4/74– 6/76	✓	✓
	6/76– 9/77	✓	✓
	9/76– 3/77	✓	✓
	1/74– 9/74		✓
	1/74– 9/74		✓
	1/74– 9/74		✓
	1/74– 6/74		✓
	12/73– 9/75	✓	✓
	12/73– 4/76	✓	✓
	12/73– 9/75	✓	✓
	2/76– 8/76		
	4/75– 5/76	✓	
	12/76– 4/77	✓	✓
	1/74– 8/74		✓
elementary school teacher	4/75– 4/77	✓	✓
personnel manager	4/75– 9/77	✓	✓
	4/75– 5/76	✓	
Montessori school teacher	12/73– 9/77	✓	✓
	4/75– 9/75		
aide, Stuyvesant Psychiatric Ward	4/75– 5/75	✓	
teacher of children with learning disabilities	9/75– 6/76	✓	
	11/74– 6/75	✓	
	5/74– 9/76	✓	✓
	4/75– 7/76	✓	✓
teacher of children with learning disabilities	9/75– 9/77	✓	✓

Teacher's Name	Age When Teacher First Worked with Nim	Student At
Karen Rosenthal	22	Brooklyn College
Dick Sanders	30	Columbia University
Bob Sapolsky	19	Harvard University
Amy Schachter	19	
Diane Schianno	18	Barnard College
Patty Sparveri	23	Columbia University School of General Studies
Kela Stevens	27	Columbia University School of General Studies
Carol Stewart	40	
Herbert Terrace	37	
Chaim Thomas	21	Columbia College
Bill Tynan	32	Columbia University
Peggy Voss	19	Barnard College
Mary Wambach	23	New York University

Occupation	Worked On Project	Taught	
		In Classroom	At Home
	4/75– 5/76	√	
	9/76– 9/77	√	√
	5/75– 9/75	√	√
animal caretaker	5/75– 9/76	√	√
at Bronx Zoo			
	4/74– 9/74		√
	9/75– 6/76	√	√
	9/74– 6/75	√	√
teacher of children	9/74– 6/75	√	√
with multiple handicaps			
professor	12/73– 9/77	√	√
	11/74– 8/75	√	√
	9/74– 9/77	√	√
	1/74– 6/74		√
	3/77– 9/77	√	√

Appendix C:
Nim's Vocabulary

Knowledge of a sign can be revealed either by the expression of the sign in the appropriate context or by appropriate behavior when the sign is presented. As discussed in Chapter 10, evidence provided by tests for sign comprehension is less persuasive about knowledge of a sign than evidence provided by demonstrations that the sign was properly expressed. As is true of children, the number of signs that Nim comprehended was considerably larger than the number of signs he expressed (see the figure on page 138 and the table on pages 166–67). Because the various photographs of Nim's signing throughout this book do not capture all of the nuances of his signs, Nim's expressive vocabulary is described in full detail in this appendix.

In order to be included in our record of Nim's expressive vocabulary, a sign had to satisfy the following criteria: it had to be observed to occur spontaneously, on separate occasions, by three different teachers, and it had to occur spontaneously at least once on five successive days. By "spontaneous," I mean that Nim was not molded to make the sign, nor was he physically prompted by his teacher, nor did the teacher make the sign in question immediately prior to Nim's expression of it.

In some cases usage, location, configuration/orientation, and/or movement changed over the period of the project. The stages are noted as follows: T = terminal stage, M = intermediate stage (where applicable), and O = original stage.

In some cases hand configurations (C) conformed to standard ASL forms and are referred to by those forms, for example, *l-*, *o-*, *s-*, and 5-hand.

Nim's Expressive Vocabulary

Acquisition Order Number	Sign	Acquisition Date	Usage
1	*drink*	3/2/74	to request or identify juices or other liquids, present or not present, in or out of eating situation
2	*up*	3/8/74	T: incorporated into *locative*. O: to indicate location of an object, or as a request to be lifted up
3	*sweet*	3/11/74	to request or identify "sweet-tasting" items (candy, honey)
4	*give*	4/1/74	to request objects out of reach and usually in possession of a teacher
5	*more*	5/15/74	used singly or in combination to request reoccurrence (continuation) of activity (*More tickle*) or additional quantity of an item (*More banana*)
6	*eat*	7/17/74	to request or identify solid foods, present or not present
7	*hug*	10/25/74	to request to be picked up and held after termination of activity or during stressful situation
8	*clean*	10/28/74	prior to or during process of washing or wiping something other than his body, cleaning implement (e.g., paper towel) often present
9	*dog*	1/25/75	for real dogs, pictures of dogs, and sometimes for bark of dog not present

Location (L)	Configuration and Orientation of Active Hand(s) (C)	Movement (M)
mouth	thumb extended up from closed fist	thumb of C to L, simple contact
at arm's length, head height	slightly curved hand, sometimes with index finger extended up from loose fist	C thrust upward from L
lower lip	index and second fingers extended from closed fist, palm toward signer	fingers (C) brush downward against L
T: in front of body O: at arm's length in front of body	flat hand, palm up	fingertips close toward palm, sometimes with bending of wrist
in front of body	T: both hands, o-configuration, thumb and forefinger side up, palm toward signer. M: same as T, only hands are loosely open. U: both hands, forefingers extended from closed fist	fingertips contact; both hands active
lips	T: o-configuration, thumb and forefinger side up, palm toward signer. O: forefinger extended from fist, or fingertips of curved hand	fingertips (C) to L, simple contact
chest and shoulders	both hands, curved toward signer	hands (C) grasp contralateral shoulders simultaneously
in front of body, flat hand, palm up	flat hand, palm down at 90-degree angle to L	C contacts L near wrist, moves along C, off fingertips
outside of thigh	loose, open hand, palm toward signer	C contacts L several times

Nim's Expressive Vocabulary

Acquisition Order Number	Sign	Acquisition Date	Usage
10	*down*	1/26/75	T: incorporated into *locative*, request to be put down or to be allowed to get down (on floor, ground, etc.). O: also to indicate position of an object, and as a request for person to lower an object
11	*open*	2/12/75	prior to opening doors and containers (boxes, jars); or to request permission to open or assistance in opening doors and containers
12	*water*	4/3/75	T: to most water (in a glass, from a sink faucet, in a river or pond, but not to water from outside lawn spigots, for which he would use *drink*). O: also to train, to indicate he had urinated in his pants
13	*listen*	4/23/75	to loud noises, ringing telephone, sound-producing objects (music box, various toys)
14	*go*	5/10/75	T: when being carried or when stationary with teacher, to initiate movement in a certain direction. O: also to request someone to move away, and to request someone push (pull) his wagon or stroller
15	*tickle*	6/5/75	to request to be tickled, sometimes before tickling a companion

Location (L)	*Configuration and Orientation of Active Hand(s) (C)*	*Movement (M)*
side or front of body	flat hand, sometimes index finger extended from closed fist, palm toward signer	extension of arm and hand in downward direction
in neutral position in front of body near waist; directly above or in front of object; on (in contact with) object	T: flat hands, palms toward object to be opened. O: palms away from signer	T: hands (C), touching at thumbs, move up and apart; may or may not contact object first; both hands active. O: hands contact object before M
lower chest, in front of body	flat hand, palm up	arm bent, fingertips (C) fold in to a closed hand, in order from little finger to forefinger; usually repeated
ear	index finger extended from closed fist, palm toward signer	fingertip (C) to L, simple contact
side and front of body	flat hand, palm down; sometimes index finger extended from loosely cupped hand	extension of arms in repeated thrusts from near body, hand sometimes flexes up and down at wrist
T: (a) back of curved hand, palm down; (b) at specific spot on body to be tickled (neck, stomach). O: (a) only	index finger extended from closed fist, palm toward L	finger (C) to L, simple contact, often repeated

Nim's Expressive Vocabulary

Acquisition Order Number	Sign	Acquisition Date	Usage
16	hand cream	6/6/75	T: to request hand cream container or that hand cream be squeezed onto his hands by the teacher. O: also request to put hand cream on his teacher
17	brush	6/14/75	to identify a brush, to request to brush himself
18	ball	6/17/75	to identify real balls and pictures of balls
19	book	6/30/75	to identify books and magazines
20	shoe	7/14/75	T: to identify real shoes and pictures of shoes, to request to take shoes off his companion. O: also to identify other things on feet, e.g., socks
21	hurt	7/16/75	to indicate an injury (e.g., scratch) on himself or a companion, or when he hurt himself (by falling and bumping his head)
22	toothbrush	7/20/75	T: to identify actual toothbrush, pictures of toothbrushes, toothpaste, or people brushing their teeth; to request his toothbrush (not present); in early evening, to request to go to his room and sleep (he always brushed his teeth just prior to going to bed for the night)
23	hurry	7/30/75	sometimes to indicate teacher was not acting quickly enough (in preparing food, opening a door)

Location (L)	Configuration and Orientation of Active Hand(s) (C)	Movement (M)
in front of chest	flat hands held palm to palm	repeated rubbing back and forth of both palms (C)
back of forearm	flat hand, fingers spread, palm down	C drawn up arm, from wrist to mid-forearm, repeatedly
in front of body	curved hands, one palm up, the other down	hands (C) contact either with fingertips of both away from signer, or with hands at 90-degree angle to each other
in front of body	flat hands, palm to palm	hands open outward in opposite directions (like a book opening), both hands active
in front of body, usually next to companion's shoes	closed fists, palms down	thumb side of fists (C) contact, move apart, contact again; often repeated several times; both hands active
at site of injury	index fingers extended from closed fists, palms down	fingertips (C) converge at L; both hands active
front of teeth, usually upper teeth	index finger extended from closed fist, palm down	side of finger (C) moves back and forth repeatedly across L
eye level, one hand on either side of head	loose, open hands, bent at wrist, palms down	vigorous shaking of both hands (C)

Nim's Expressive Vocabulary

Acquisition Order Number	Sign	Acquisition Date	Usage
24	come	8/2/75	T: to request someone to approach. O: also sometimes for inanimate objects (ball, wagon) out of reach
25	harmonica	8/25/75	to identify picture of harmonica; to request harmonica, present
26	red	8/26/75	for red objects; usually in response to question about object's color; sometimes, to identify an apple, whether red, yellow, or green
27	play	8/29/75	T: (a) to request companion to chase him; (b) sometimes as request for other animals (cow, dog) to engage in chase game. O: (a) only
28	me	9/13/75	to identify himself as the current or future recipient of an object or food or participant in an activity (play, jump)
29	banana	10/5/75	to identify real bananas, plastic bananas, and pictures of bananas; to request real banana
30	gum	10/13/75	T: to request gum. O: also sometimes for candy with gumlike consistency
31	hat	10/16/75	T: to identify real hats, pictures of hats; to request real hat. O: also for articles of clothing placed on the head

Location (L)	*Configuration and Orientation of Active Hand(s) (C)*	*Movement (M)*
in front of body	T: flat hand, palm to the side. O: palm up	extended arm, bends slightly at elbow; hand (C) bends at wrist in beckoning motion
closed lips	index finger extended from closed fist, side of finger toward signer	C placed between L; signer then blows forcefully against finger
lower lip	index finger extended up from closed fist, palm toward signer	T: finger (C) contacts and flips L in downward motion. O: simple contact
in front of body	flat hands, palms separated, facing each other	T: hands (C) clap together, often repeatedly, sometimes only once; single clap often loud and signed when companion not looking, to attract attention. O: vigorous clapping; both hands active
center chest	T: index finger extended from closed fist. O: flat hand, palm in toward signer; sometimes as in T, above	simple contact, often repeated
in front of body, on index finger extended from closed fist of passive hand	index finger extended from closed fist, palm toward signer	T: C index finger moves from tip to base of L, index finger along top side; usually repeated several times. O: C moves along one side, then other side of L
cheek	index and second fingers extended from closed fist, palm down	fingers (C) contact L, extended fingers bend into cheek and straighten again; often repeated
top of head	flat hand, palm down	C to L, repeated contact

Nim's Expressive Vocabulary

Acquisition Order Number	Sign	Acquisition Date	Usage
32	*apple*	10/20/75	to identify real apples, plastic apples, pictures of apples; to request real apples, or apple-tasting foods (apple juice, apple sauce)
33	*groom*	10/20/75	to request his companion groom him
34	*Nim*	11/26/75	to identify himself as the current or future recipient of an object or food or participant in an activity (play, jump)
35	*key*	12/2/75	to identify real keys and pictures of keys
36	*sorry*	12/7/75	after Nim has done something for which he is being or usually will be reprimanded; when companion is emotionally upset, not necessarily because of Nim
37	*orange*	12/7/75	T: (a) to identify real oranges, plastic oranges, pictures of oranges; (b) to identify color of orange objects. O: (a) only
38	*tea*	12/12/75	to identify tea bags, cup of tea; to request tea; in the presence of and to pictures of a type of cup in which he usually received tea, no tea present
39	*nut*	12/15/75	to identify or request any of a variety of nuts
40	*raisin*	12/18/75	to identify or request raisins

Location (L)	Configuration and Orientation of Active Hand(s) (C)	Movement (M)
cheek	T: index finger extended from closed fist but bent at knuckle, palm down. O: index and second fingers extended	finger (C) on L, rotates clockwise
head, leg, or, usually, arm	o-shaped hand, palm toward body locus of sign	thumb and fingers (C) open and close repeatedly on body hair in grasping (grooming) motions on L
top/side of head	T: index and second fingers extended from a closed fist, palm toward signer. O: index finger extended	fingers (C) contact L, drawn down side of head toward ear
in front of body, on palm of flat hand, facing to side	index and second fingers extended from closed fist but bent at knuckle, palm down	fingers (C) on L; rotates clockwise
chest	closed o-hand, palm down	C on L; repeated circular motion
closed mouth	T: closed fist, thumb side toward mouth, palm down. O: fist touches mouth	T: C contacts L, usually repeatedly. O: fist turns clockwise from wrist
o-hand, thumb side up	T: index finger extended from closed fist. O: also sometimes index finger extended from spread fingers of flat hand	finger (C) inserted down into opening in L
upper front teeth	thumb extended up from closed fist	thumbnail of C placed behind L, flicked outward, normally creating a clicking sound
cheek	middle finger protrudes forward from 5-hand, palm toward face	T: finger (C) flicks downward off L. O: C flicks upward

Nim's Expressive Vocabulary

Acquisition Order Number	Sign	Acquisition Date	Usage
41	*smell*	1/27/76	T: in presence of strong, usually sweet-scented items (perfume, flowers). O: for variety of strong smells (perfume, gasoline fumes, after he defecated)
42	*pants*	1/28/76	T: to identify his pants; to request to put on or take off his pants. O: to request that his pants be put on or taken off
43	*you*	1/29/76	to indicate the other participant (his companion) in a sign exchange
44	*bug*	2/3/76	to identify a variety of insects (flies, ants)
45	*hot*	2/4/76	in the presence of heat from a specific source (floodlights for filming, hotplate, flames, hot water, tea)
46	*in*	2/13/76	to indicate location of objects inside containers; prior to putting on his shirt or pants; to request to put on his companion's shoe(s) or to go inside the house
47	*powder*	2/17/76	to identify container of baby powder; to request baby powder be put on him, usually during diaper change
48	*out*	2/26/76	T: (a) prior to removing objects from containers, sometimes to indicate a need for assistance in the task; (b) prior to removing his clothes; (c) to request to remove his companion's shoe(s);

Location (L)	Configuration and Orientation of Active Hand(s) (C)	Movement (M)
nose	T: flat hand, palm toward signer. O: closed fist, index finger extended	T: C moves in and out near nostrils repeatedly. O: finger (C) to L, simple contact
both legs	flat hands, palms toward signer	C to lower leg, drawn up to knee; both hands active
center of companion's chest; from a distance, at arm's length in front of body	index finger extended from closed fist	finger (C) to L; simple contact C pointed in toward companion
nose	thumb, index finger, and third finger extended from loosely closed hand	T: C to L, simple contact of thumb. O: index and third fingers curve inward twice after contact
near face	loose flat hand, palm toward signer	C moves repeatedly back and forth vigorously at L in fanning motion
c-hand, thumb side up, palm angled toward signer	loose flat hand, palm toward signer	C inserted downward between thumb and palm of L
inside of thigh	flat hand, palm toward signer	C to L; repeated contact
c-hand, thumb side up, palm angled toward signer	loose flat hand, palm toward signer	C inserted downward between thumb and palm of L, then withdrawn quickly

Nim's Expressive Vocabulary

Acquisition Order Number	Sign	Acquisition Date	Usage
			(d) to request to leave a location where a door must be opened first (classroom, car, house). O: (a) and (d) only
49	*that/there* (later redesignated as *point*)	2/27/76	T: (a) to draw attention to a specific object by pointing at it; (b) to indicate the location of an object; (c) to indicate the direction/location of where he wanted his companion and him to go. O: (a) and (b) only
50	*please*	2/27/76	in request situations, prior to sign indicating request (*Please hug, Please play*)
51	*flower*	3/9/76	to identify real flowers, plastic flowers, pictures of flowers
52	*Laura*	3/12/76	to identify Laura Petitto, a teacher
53	*kiss*	3/12/76	to request a kiss; prior to kissing someone
54	*cracker*	3/14/76	to identify crackers, breads
55	*light*	3/16/76	to identify bright sources of light (floodlights, occasionally the sun, beams of light from a flashlight or sunlight highlighting dust in the air in dim room)
56	*jump*	3/31/76	prior to jumping; to request teacher to join in jump-chase game with him
57	*rock*	4/7/76	to identify rocks

Location (L)	Configuration and Orientation of Active Hand(s) (C)	Movement (M)
at arm's length, in front of body	index finger extended from closed fist, palm in various orientations	finger (C) points at object, direction, place, etc.
center chest	flat hand, palm toward signer	C to L, moves in circular motion
nose	curved hand, palm toward signer	C to the side of L, simple contact
eyebrow	T: *l* hand.* O: index finger extended from closed fist	T: thumb of C contacts L. O: tip of index finger contacts L
protruding lips	flat hand, palm toward signer	C to L, kisses center palm
elbow of bent, upward extended arm	loose fist	C taps L repeatedly
at arm's length, in front of body	closed fist, palm down; occasionally palm up toward light sources especially when first learning sign	C opens to a 5-hand
flat hand, palm up or angled up	index and third fingers extended from closed fist; often index finger only	fingers (C) to L bend on contact, then straighten, as if "jumping" off palm
back of closed fist, palm down	closed fist, index and third fingers extended slightly and bent at knuckle, palm up	C taps L twice, sometimes repeatedly

* Change in configuration due to teacher altering name sign

Nim's Expressive Vocabulary

Acquisition Order Number	Sign	Acquisition Date	Usage
58	*work*	4/13/76	prior to or during structured activities
59	*Andrea*	4/17/76	to identify Andrea Liebert, a teacher
60	*bite*	4/17/76	in a stressful situation in which his behavior and facial expression are similar to when he may bite his companion; when first learning the sign, he often would not bite his companion if he had signed *bite* first
61	*chair*	4/24/76	to identify a variety of chairs (high chair, car seat, benches, potty); to request teacher to push him in roller-type office chair
62	*pole*	4/28/76	to identify poles; prior to climbing a pole
63	*dirty*	4/30/76	T: (a) to indicate he has to urinate or defecate; (b) to terminate the current structured activity even though he does not have to use the toilet. O: (a), and to indicate soiled items
64	*spoon*	5/3/76	to identify real spoons and pictures of spoons
65	*happy*	5/10/76	when excited, as in tickle game
66	*Walter*	5/11/76	to identify Walter Benesch, a teacher

Location (L)	Configuration and Orientation of Active Hand(s) (C)	Movement (M)
T: inside of wrist, closed fist, palm down. O: back of closed fist, palm down	T: closed fist, palm angled down. O: palm down	T: C contacts L repeatedly. O: C contacts L twice
ear	closed fist, palm toward signer	C to contralateral L; contact often repeated
flat hand, palm down	loose c-hand, palm down	C clamps on L, grabbing meat of palm
open c-hand, palm angled away from signer	index and middle fingers extended from loosely closed fist	fingers (C) hand over thumb (L)
in front of body	closed fist, thumb side up, palm toward signer	bottom of C contacts top of L, then C moves upward
underside of chin	back of loose open hand, palm down, bent at wrist	C to L, repeated contact
flat hand, palm up	index and third fingers, often only index finger, extended from closed fist, angled sideways	fingers (C) contact L, continue upward slightly in "spooning" motion
upper chest, near shoulders	flat hands, palms toward signer	C contacts L in repeated vigorous upward brushing strokes
shoulder	flat hand, palm toward signer	C taps contralateral L

Nim's Expressive Vocabulary

Acquisition Order Number	Sign	Acquisition Date	Usage
67	*angry*	5/12/76	in stressful situations when prior requests have been denied by companion, who often is angry at him
68	*finish*	5/28/76	to indicate transition from ongoing activity
69	*hungry*	6/5/76	usually when he has not eaten recently; food may or may not be present
70	*Herb*	6/10/76	to identify Dr. Herbert Terrace, principal investigator and a teacher
71	*cat*	6/29/76	to identify real cats and pictures of cats
72	*pear*	7/7/76	to identify real pears, plastic pears, and pictures of pears
73	*brown*	7/28/76	to identify brown objects
74	*bird*	7/29/76	to identify real birds and pictures of birds
75	*grape*	7/31/76	to identify real grapes, plastic grapes, and pictures of grapes; sometimes to identify grape-shaped food for which he did not have a specific sign
76	*berry*	8/17/76	to identify various berries and berrylike fruit (cherries, strawberries, blueberries, wild berries)
77	*fruit*	8/18/76	to identify less common fruit for which he has signs (peach);

Location (L)	Configuration and Orientation of Active Hand(s) (C)	Movement (M)
face	loose hand, fingertips up, palm toward signer, ends in closed fist	C begins at top of L, moves down off L as hand closes
in front of body	loose hands, palms up, fingers pointing away from signer	hands touch, rotate inward, and move apart until palms face downward
center chest	curved hand, palm toward signer	C to base of neck, down L; sometimes repeated
upper lip	index and third fingers extended from closed fist, palm toward signer	fingers (C) to L, stroking downward and away from mouth twice
upper lip	index fingers extended from closed fists, palms toward signer	fingers (C) contact L, drawn outward from face (mimics cat whiskers); both hands active
in front of body, all fingertips touching and directed upward	c-hand, palm facing L	C moves down over, closes around L, is drawn upward and off L
side of face	flat hand, palm toward signer	C contacts L at temple, moves downward
mouth	index finger and thumb extended from closed fist, palm facing signer; sometimes facing away from signer	tips of index finger and thumb (C) touch in front of L
in front of body, flat hand, palm up	index finger and thumb extended from closed fist, palm facing L	fingers (C) contact L with "picking" motion, then move away from L; usually repeated
5-hand, angled toward signer	loose fist, thumb toward L	C grasps little finger of L, thumb end first
cheek	f-hand, palm facing signer	tips of index finger and thumb (C) contact L

Nim's Expressive Vocabulary

Acquisition Order Number	Sign	Acquisition Date	Usage
			sometimes to identify fruit for which he does not have a specific sign (cantaloupe)
78	*help*	9/24/76	to request assistance from his companion in doing a task
79	*bad*	9/27/76	after a teacher has scolded him for misbehaving, to acknowledge that he has misbehaved
80	*baby*	10/6/76	to identify or request any doll-like object (baby doll, stuffed animal)
81	*wash*	10/21/76	to request to wash his hands in sink either by turning faucet on or in water already in sink
82	*yogurt*	11/23/76	to identify yogurt, usually in 8-oz. cups; to request yogurt, not present
83	*sleep*	1/4/77	in his room just before going to bed for the night; in late afternoon after an active day; during a structured activity in which he seems to have little interest
84	*pull*	1/28/77	to request pulling games (teacher pulling on his leg), or to be pulled in a wagon
85	*blue*	2/4/77	to identify blue objects, especially blue "cookie monster" doll
86	*black*	3/16/77	to identify black objects
87	*paper*	3/16/77	to identify a variety of paper, except napkins and paper towels
88	*Joyce*	3/29/77	to identify Joyce Butler, a teacher

Location (L)	Configuration and Orientation of Active Hand(s) (C)	Movement (M)
in front of body, flat hand, palm up	closed fist, thumb side up	C, resting on L, moves upward both hands active
mouth	loose flat hand, palm facing signer	C touches or passes in front of L in downward direction; slight rotation of palm downward
in front of chest	both hands grasp opposite elbows firmly	no movement; occasionally slight rocking motion
in front of body	cupped hands, palms facing	hands (C) clasp, rub each other in washing motion
in front of chest	loose first, thumb side up	C closes and opens repeatedly in "milking" motion
temple	flat hands, palms touching, fingertips up	back of one hand (C) contacts L
in front of body	closed fists, palms up, bottom of one fist touching top of the other; both fists parallel to the ground	C moves toward signer; usually repeated
stomach	flat hand, palm toward signer, fingertips to side	C contacts L; usually repeated
eyebrow or eyebrow ridge	loose fist, index finger extended, palm toward signer	finger (C) contacts L, drawn along L toward side of head
in front of body	loose flat hands, palms facing, fingertips in opposite directions; right hand often in loose fist	heels of C contact, usually twice; both hands active
wrist; loose flat hand, palm angled toward signer	c-hand, palm toward back of L wrist	C grasps dorsal side of L

Nim's Expressive Vocabulary

Acquisition Order Number	Sign	Acquisition Date	Usage
89	*Bill*	3/31/77	to identify Bill Tynan, a teacher
90	*house*	3/31/77	to identify pictures of houses; at the end of a classroom day at Columbia when he was preparing to be driven home; sometimes when first saw his house after driving back from Columbia, but not to other houses
91	*Susan*	4/4/77	to identify Susan Quinby, a teacher
92	*thirsty*	4/15/77	to request a drink, which may or may not be present
93	*tree*	4/17/77	to identify trees or drawings of trees
94	*Dick*	4/21/77	to identify Dick Sanders, a teacher
95	*lie down*	4/25/77	to request to lie down in teacher's lap or on the floor or ground
96	*cut*	5/4/77	during activities requiring cutting with scissors or a knife
97	*napkin*	5/11/77	to identify napkins and paper towels used for wiping face and mouth
98	*balloon*	5/27/77	to identify balloon and pictures of balloons
99	*cookie*	6/1/77	to identify cookies and cakes
100	*Renee*	6/3/77	to identify Renee Falitz, a teacher

Location (L)	*Configuration and Orientation of Active Hand(s) (C)*	*Movement (M)*
ear	closed fist, palm toward signer	C contacts L
in front of chest	flat hands, palms facing, fingertips pointing up or away from signer	fingertips (C) touch; both hands active
upper arm	s-fist, palm toward signer	C contacts contralateral L; usually repeated
throat	closed fist, index finger extended	finger (C) contacts L, moves downward
loose 5-hand, palm up, arm across front of body	arm, bent at elbow, pointing up; loose open hand, palm toward signer	elbow (C) rests on L; C rotates back and forth
side of upper chest	closed fist, index finger extended	finger (C) contacts contralateral L, moves downward diagonally
in front of body, flat hand, palm up	loose fist, index and third fingers extended and slightly bent, palm up	back of fingers (C) contact L
in front of body, edge or inside of wrist; arm horizontal	flat hand, palm toward signer	little finger side of C touches and crosses over L; sometimes C moves up arm instead of across
mouth	loose fist, thumb extended, palm toward signer	side of thumb (C) contacts corner of L, moves across lips
corners of mouth	flat o-hands, palms down	thumb and index finger of each hand contacts L, simultaneously move outward slightly
in front of body, flat hand, palm up	claw hand, palm down	fingertips of C contact L
back of neck	flat hand, palm toward signer	C contacts L

Nim's Expressive Vocabulary

Acquisition Order Number	Sign	Acquisition Date	Usage
101	*cup*	6/11/77	to identify cups
102	*bowl*	6/13/77	to identify bowls
103	*Steve*	6/19/77	to identify Steve Carey, a teacher
104	*ice*	6/19/77	to identify ice and popsicles
105	*run*	6/20/77	to request his companion to engage in a run-chase game
106	*climb*	6/21/77	to request to climb
107	*Mary*	6/26/77	to identify Mary Wambach, a teacher
108	*box*	7/4/77	to identify all square and rectangular containers, regardless of size
109	*ear*	8/9/77	to identify ears of animate objects (humans, cats) and of dolls
110	*eye*	8/9/77	to identify eyes of animate objects and of dolls
111	*nose*	8/9/77	to identify noses of animate objects and of dolls
112	*shirt*	8/10/77	to identify all pullover-type shirts worn by him or his doll

Location (L)	*Configuration and Orientation of Active Hand(s) (C)*	*Movement (M)*
in front of body, flat hand, palm up	*c*-hand, thumb side up, palm toward signer	little finger side of C contacts L
in front of stomach	cupped hands, palms up	little finger edges of C touch; both hands active
forehead	5-hand, fingertips up, thumb toward signer	thumb (C) contacts L
between teeth	closed fist, index and third finger extended and bent at knuckles, little finger side up, palm away from signer	finger of C in L
in front of body	*l*-hand, thumb up, index finger pointing away from signer	index finger (C) hooks around base of thumb (L)
on the object when near the object to be climbed; flat hand, palm angled up and toward signer when away from object	closed fist, index and middle fingers extended and slightly bent	side of third finger (C) contacts L
cheek	claw hand, fingers straight, thumb side up, fingertips toward signer	fingertips (C) contact L, move back toward ear, then away
in front of chest	flat hands, thumb side up, fingertips away from signer, several inches separating palms	C moves down in arc, hands representing sides of box
ears	*a*-fists, thumbs slightly separated from fists and toward signer, palms forward	C grasps L between thumb and index finger; both hands active
eyes	closed fist, index finger extended	fingertips (C) contact L; both hands active
nose	closed fist, index finger extended	C to L, simple contact
chest	closed fists, thumb and index finger extended	C contacts L, pinches shirt, pulls away from signer,

Nim's Expressive Vocabulary

Acquisition Order Number	Sign	Acquisition Date	Usage
113	teeth	8/27/77	to identify teeth of animate objects and dolls
114	throw	8/27/77	to request his companion to throw something (e.g., a ball) to him
115	fish	8/27/77	to identify goldfish in the pond at his home
116	wagon	8/29/77	to identify his wagon
117	glasses	9/4/77	to identify eyeglasses
118	goodbye	9/6/77	to his former teacher following a change in teachers; sometimes to other people leaving his presence
119	music	9/10/77	to identify music emanating from music boxes and musical toys
120	walk	9/17/77	prior to or during walk with teacher
121	green	9/19/77	to identify green objects
122	pool	9/19/77	to identify his plastic wading pool
123	Bob	9/19/77	to identify Bob Johnson, a teacher

Location (L)	*Configuration and Orientation of Active Hand(s) (C)*	*Movement (M)*
	toward signer, thumb side up	letting go of shirt; both hands active
teeth	closed fist, index finger extended toward signer, palm down	fingertip (C) contacts L
off side of head	open hand, palm away from signer; arm bent at elbow and raised	C moves forward from L in throwing motion
in front of body	flat hands, palms facing, thumb sides up	fingertips of one hand contact heel of other; both hands move back and forth horizontally in swimming motion
in front of body, flat hand, palm down	closed fist, thumb side up, palm toward signer	little finger (C) rests on L as L moves diagonally across space in front of signer
eyes	flat *o*-hands, palms down	thumb and index finger side of C contacts L; signer looks through *o* of hands
off side of head	flat hand, palm forward, arm bent at elbow and raised	C moves up and down at wrist
inside of forearm; arm bent at elbow, flat hand angled up and toward signer	loose flat hand, palm toward signer	C moves back and forth from elbow to wrist (L)
in front of chest; on ground	flat hands, palms down	hands (C) move forward successively in walking motion
stomach	closed fist, thumb and index finger extended toward side	thumb (C) contacts L
in front of chest	5-hands, palms toward signer, arms forming circle	fingers (C) interlace to form closed circle
inside of elbow; arm bent, flat hand, palm up	flat hand, palm down	C contacts L

Nim's Expressive Vocabulary

Acquisition Order Number	Sign	Acquisition Date	Usage
124	*hello*	9/21/77	upon arrival of another person
125	*peach*	9/21/77	to identify peaches

Location (L)	*Configuration and Orientation of Active Hand(s) (C)*	*Movement (M)*
temple area	flat hand, palm forward, arm bent at elbow and raised	index finger edge of hand (C) contacts L, moves out and away
cheek	claw hand, fingers straight and pointing toward L	C contacts L as fingertips are drawn together; C moves away from L

Index

About the Author

Herbert S. Terrace is Professor of Psychology at Columbia University. He gained his bachelor's and master's degrees from Cornell University and his doctorate from Harvard, where he studied with noted behaviorist B. F. Skinner. From 1973 to 1975 he served as President of the Society for the Experimental Analysis of Behavior. He is co-author of *Introduction to Statistics* (with Scott Parker) and *Psychology and Human Behavior* (with T. G. Bever), and is a contributor to a number of scientific journals.

A Note on the Type

This book was set in Caledonia, a Linotype face designed by W. A. Dwiggins. It belongs to the family of printing types called "modern face" by printers—a term used to mark the change in style of type letters that occurred about 1800. Caledonia borders on the general design of Scotch Modern, but is more freely drawn than that letter.

Composed by Maryland Linotype Composition Company, Inc., Baltimore, Maryland. Printed and bound by The Murray Printing Company, Forge Village, Massachusetts.

Typography and binding design by Virginia Tan